Soil Testing: Prospects for Improving Nutrient Recommendations

Related Society Publications

Soil Testing and Plant Analysis, Third Edition

Soil Testing: Correlating and Interpreting the Analytical Results

Soil Testing: Sampling, Correlation, Calibration, and Interpretation

For information on these titles, please contact the SSSA Headquarters Office; Attn: Marketing; 677 South Segoe Road; Madison, WI 53711-1086. Phone: (608) 273-8080. Fax: (608) 273-2021.

Soil Testing: Prospects for Improving Nutrient Recommendations

Proceedings of symposiums sponsored by S-3, S-4, S-8, and S-9 of the American Society of Agronomy and Soil Science Society of America. The papers were presented during the annual meetings in Cincinnati, OH, 7–12 Nov. 1993.

Co-Editors
John L. Havlin
Jeffrey S. Jacobsen

Editorial Committee
John L. Havlin
Jeffrey S. Jacobsen
Dale F. Leikam
Paul E. Fixen
Gary W. Hergert

Symposium Planning Committee
John L. Havlin
Jeffrey S. Jacobsen
Charles W. Rice
Gary M. Pierzynski

Editor-in-Chief
Larry P. Wilding

Managing Editor
David M. Kral

Associate Editor
Marian K. Viney

SSSA Special Publication Number **40**

**Soil Science Society of America, Inc.
American Society of Agronomy, Inc.
Madison, Wisconsin, USA**
1994

Cover Design: Patricia Scullion

Copyright © 1994 by the Soil Science Society of America, Inc. and the American Society of America, Inc.

ALL RIGHTS RESERVED UNDER THE U.S. COPYRIGHT LAW OF 1976 (P.L. 94-553)

Any and all uses beyond the limitations of the "fair use" provision of the law require written permission from the publisher(s) and/or the author(s); not applicable to contributions prepared by officers or employees of the U.S. Government as part of their official duties.

Soil Science Society of America, Inc.
American Society of America, Inc.
677 South Segoe Road, Madison, WI 53711 USA

Library of Congress Cataloging-in-Publication Data
Soil testing : prospects for improving nutrient recommendations / co
 -editors, John L. Havlin, Jeffrey S. Jacobsen ; editorial committee,
John L. Havlin . . . [et al.] ; symposium planning committee, John L.
Havlin . . . [et al.] ; managing editor, D.M. Kral ; associate editor,
M.K. Viney.
   p.   cm. — (SSSA special publication ; no. 40)
  "Proceedings of symposiums sponsored by S-4, S-8, S-3, and S-9 of
the American Society of Agronomy and Soil Science Society of
America. The papers were presented during the annual meetings in
Cincinnati, OH, 7-12 Nov. 1993."
  Includes bibliographical references and index.
  ISBN 0-89118-815-0
  1. Soils—Analysis—Congresses. 2. Soil fertility—Congresses.
3. Fertilizers—Congresses. I. Havlin, John. II. Jacobsen,
Jeffrey S. III. American Society of Agronomy. Division S-4.
IV. Series.
S593.S7428   1994
631.4'22'0287—dc20   94-31653
                                                                                            CIP

Printed in the United States of America

CONTENTS

Foreword vii

Preface ix

Contributors xiii

Conversion Factors for SI and non-SI units xv

1 Integrating Mineralizable Nitrogen Indices into Fertilizer Nitrogen Recommendations
   Charles W. Rice and John L. Havlin 1

2 Potential Nitrogen Mineralization: Laboratory and Field Evaluation
   M. L. Cabrera, M. F. Vigil, and D. E. Kissel 15

3 Field Indicators of Nitrogen Mineralization
   J. S. Schepers and J. J. Meisinger 31

4 Linking Nitrogen Mineralization and Plant Nitrogen Demand with Thermal Units
   C. Wayne Honeycutt 49

5 Evaluating Potential Nitrogen Mineralization for Predicting Fertilizer Nitrogen Requirements of Long-Term Field Experiments
   C. A. Campbell, Y. W. Jame, O. O. Akinremi, and H. J. Beckie 81

6 Current Phosphorus Availability Indices: Characteristics and Shortcomings
   F. R. Cox 101

7 Innovative Soil Phosphorus Availability Indices: Assessing Inorganic Phosphorus
   Andrew N. Sharpley, J. T. Sims, and Gary M. Pierzynski ... 115

8 Innovative Soil Phosphorus Availability Indices: Assessing Organic Phosphorus
   H. Tiessen, J. W. B. Stewart, and A. Oberson 143

9 Site-Specific Soil Tests and Interpretations for Potassium
   D. J. Eckert 163

10 Effects of Iron Oxidation State on the Fate and Behavior of Potassium in Soils
   Siyuan Shen and Joseph W. Stucki 173

11 Reinventing Soil Testing for the Future
   Earl O. Skogley 187

12 Current Approaches to Soil Testing Methods: Problems and Solutions
   E. R. Allen, L. G. Unruh, and G. V. Johnson 203

FOREWORD

Considerable progress has been made in the past 40 years to develop and implement modern methods of plant nutrient analysis in soils. This has facilitated enhanced crop production, economic viability for producers and minimized hazards of water quality pollution from excess nutrients. Increased emphasis will be placed on nutrient availability in soils as precision farming techniques are further developed and implemented. This publication focuses heavily on nitrogen availability and mineralization from organic matter. One of the difficulties in developing recommendations from nitrogen soil tests is our inability to accurately predict the rate of nitrogen mineralization. Hence, soil tests are not exact measures of available nutrients because of the difficulty in using a simple chemical extractant to evaluate the available nutrients for plants in a wide variety of soils with extremely diverse mineraology. Furthermore, those nutrients, such as nitrogen and phosphorous that cycle through both organic and inorganic forms are difficult to measure accurately because the conversion between forms depends on climate, tillage management, cropping history, and soil fertility. In spite of these complexities, considerable progress has been made in improving our understanding and ability to predict nutrient availability under complex conditions. The purpose of this publication is to bring together the latest ideas and knowledge for improving nutrient recommendations. The co-editors, editorial committee, and symposium planning committee are to be commended for bringing this important text to fruition. It will stimulate further ideas and research to improve nutrient recommendations. This takes on added meaning at this time when the public mandates increased attention to the sustainability of the biosphere.

Larry P. Wilding, *President*
Soil Science Society of America

PREFACE

Agriculture production has been implicated a major source of surface water and groundwater contamination by plant nutrients. As environmental concerns related to nutrient use and management increase, soil testing will be relied on to help identify the most profitable and environmentally stable nutrient application rate. Recent research in soil testing for nitrogen, phosphorus, and potassium have demonstrated the potential for improving the accuracy of predicting optimum nutrient rates, and therefore, reduce the potential for environmental contamination.

Four symposia were held at the 1993 American Society of Agronomy meeting in Cincinnati, OH, to address the prospects of improving nitrogen, phosphorus, and potassium recommendations through alternative soil testing approaches. This publication contains the papers presented at the symposia. It is our intention that this publication stimulate the agricultural science and industry community to expand research and demonstration of alternative soil testing strategies for improving the accuracy of nutrient recommendations. Clearly, these and/or other methods that ultimately improve the reliability of the soil testing-nutrient recommendation system will help to reduce agricultures impact on the environment.

The Editorial Committee
J. L. HAVLIN, co-chair
Kansas State University
Manhattan, Kansas

J. S. JACOBSEN, co-chair
Montana State University
Bozeman, Montana

D. F. LEIKAM
Farmland Industries
Manhattan, Kansas

P. E. FIXEN
Potash & Phosphate Institute
Brookings, South Dakota

G. W. HERGERT
University of Nebraska
North Platte, Nebraska

CONTRIBUTORS

O. O. Akinremi	Research Associate, Agriculture and Agri-Food Canada, Swift Current, SK S9H 3X2
E. R. Allen	Assistant Professor, Department of Agronomy, Oklahoma State University, Stillwater, OK 74078
H. J. Beckie	Research Scientist, Agriculture and Agri-Food Canada, Melfort, SK S0E 1A0
M. L. Cabrera	Assistant Professor of Crop and Soil Sciences, Crop and Soil Sciences Department, University of Georgia, Athens, GA 30602
C. A. Campbell	Principal Research Scientist, Agriculture and Agri-Food Canada, Swift Current, SK S9H 3X2
F. R. Cox	Professor, Soil Science Department, North Carolina State University, Raleigh, NC 27695-7619
D. J. Eckert	Professor of Soil Fertility, Department of Agronomy, Ohio State University, Columbus, OH 43210
John L. Havlin	Associate Professor, Department of Agronomy, Kansas State University, Manhattan, KS 66506-5501
C. Wayne Honeycutt	Soil Scientist, USDA-ARS, New England Plant, Soil, and Water Laboratory, University of Maine, Orono, ME 04469
Y. W. Jame	Research Scientist, Agriculture and Agri-Food Canada, Swift Current, SK S9H 3X2
G. V. Johnson	Professor, Department of Agronomy, Oklahoma State University, Stillwater, OK 74078
D. E. Kissel	Professor and Head, Department of Crop and Soil Sciences, University of Georgia, Athens, GA 30602-7272
J. J. Meisinger	Soil Scientist, USDA-ARS, BARC-West, Beltsville, MD 20705
A. Oberson	Post Doctorate Fellow, College of Agriculture, Tropical Pastures Program, CIAT, Cali, Colombia
Gary M. Pierzynski	Associate Professor, Department of Agronomy, Kansas State University, Manhattan, KS 66506
Charles W. Rice	Associate Professor, Department of Agronomy, Kansas State University, Manhattan, KS 66506-5501

J. S. Schepers	Soil Scientist, USDA-ARS, University of Nebraska, Lincoln, NE 68583-0934
Andrew N. Sharpley	Soil Scientist, USDA-ARS, National Agricultural Water Quality Laboratory, Durant, OK 74702-1430
Siyuan Shen	Graduate Research Assistant, Department of Agronomy, University of Illinois, Urbana, IL 61801
J. T. Sims	Professor of Soil Science, Department of Plant and Soil Sciences, University of Delaware, Newark, DE 19717-1303
Earl O. Skogley	Professor of Soil Management and Fertility, Plant, Soil, and Environmental Sciences Department, Montana State University, Bozeman, MT 59717
J. W. B. Stewart	Professor of Soil Science, College of Agriculture, University of Saskatchewan, Saskatoon, SK S7N 0W0 Canada
Joseph W. Stucki	Professor of Soil Science, Agronomy Department, University of Illinois, Urbana, IL 61801
H. Tiessen	Adjunct Professor of Soil Science, College of Agriculture, University of Saskatchewan, Saskatoon, SK S7N 0W0 Canada
L. G. Unruh	Extension Soil Chemistry, Department of Soil and Crop Sciences, Texas A&M University, College Station, TX 77843
M. F. Vigil	Soil Scientist, USDA-ARS, Central Great Plains Research Station, Akron, CO 80720

Conversion Factors for SI and non-SI Units

Conversion Factors for SI and non-SI Units

To convert Column 1 into Column 2, multiply by	Column 1 SI Unit	Column 2 non-SI unit	To convert Column 2 into Column 1, multiply by
		Length	
0.621	kilometer, km (10^3 m)	mile, mi	1.609
1.094	meter, m	yard, yd	0.914
3.28	meter, m	foot, ft	0.304
1.0	micrometer, μm (10^{-6} m)	micron, μ	1.0
3.94×10^{-2}	millimeter, mm (10^{-3} m)	inch, in	25.4
10	nanometer, nm (10^{-9} m)	Angstrom, Å	0.1
		Area	
2.47	hectare, ha	acre	0.405
247	square kilometer, km^2 (10^3 m)^2	acre	4.05×10^{-3}
0.386	square kilometer, km^2 (10^3 m)^2	square mile, mi^2	2.590
2.47×10^{-4}	square meter, m^2	acre	4.05×10^3
10.76	square meter, m^2	square foot, ft^2	9.29×10^{-2}
1.55×10^{-3}	square millimeter, mm^2 (10^{-3} m)^2	square inch, in^2	645
		Volume	
9.73×10^{-3}	cubic meter, m^3	acre-inch	102.8
35.3	cubic meter, m^3	cubic foot, ft^3	2.83×10^{-2}
6.10×10^4	cubic meter, m^3	cubic inch, in^3	1.64×10^{-5}
2.84×10^{-2}	liter, L (10^{-3} m^3)	bushel, bu	35.24
1.057	liter, L (10^{-3} m^3)	quart (liquid), qt	0.946
3.53×10^{-2}	liter, L (10^{-3} m^3)	cubic foot, ft^3	28.3
0.265	liter, L (10^{-3} m^3)	gallon	3.78
33.78	liter, L (10^{-3} m^3)	ounce (fluid), oz	2.96×10^{-2}
2.11	liter, L (10^{-3} m^3)	pint (fluid), pt	0.473

CONVERSION FACTORS FOR SI AND NON–SI UNITS

To convert Column 1 into Column 2, multiply by	Column 1 SI Unit	Column 2 non-SI Unit	To convert Column 2 into Column 1, multiply by
	Mass		
2.20×10^{-3}	gram, g (10^{-3} kg)	pound, lb	454
3.52×10^{-3}	gram, g (10^{-3} kg)	ounce (avdp), oz	28.4
2.205	kilogram, kg	pound, lb	0.454
0.01	kilogram, kg	quintal (metric), q	100
1.10×10^{-3}	kilogram, kg	ton (2000 lb), ton	907
1.102	megagram, Mg (tonne)	ton (U.S.), ton	0.907
1.102	tonne, t	ton (U.S.), ton	0.907
	Yield and Rate		
0.893	kilogram per hectare, kg ha^{-1}	pound per acre, lb acre^{-1}	1.12
7.77×10^{-2}	kilogram per cubic meter, kg m^{-3}	pound per bushel, bu^{-1}	12.87
1.49×10^{-2}	kilogram per hectare, kg ha^{-1}	bushel per acre, 60 lb	67.19
1.59×10^{-2}	kilogram per hectare, kg ha^{-1}	bushel per acre, 56 lb	62.71
1.86×10^{-2}	kilogram per hectare, kg ha^{-1}	bushel per acre, 48 lb	53.75
0.107	liter per hectare, L ha^{-1}	gallon per acre	9.35
893	tonnes per hectare, t ha^{-1}	pound per acre, lb acre^{-1}	1.12×10^{-3}
893	megagram per hectare, Mg ha^{-1}	pound per acre, lb acre^{-1}	1.12×10^{-3}
0.446	megagram per hectare, Mg ha^{-1}	ton (2000 lb) per acre, ton acre^{-1}	2.24
2.24	meter per second, m s^{-1}	mile per hour	0.447
	Specific Surface		
10	square meter per kilogram, m^2 kg^{-1}	square centimeter per gram, cm^2 g^{-1}	0.1
1000	square meter per kilogram, m^2 kg^{-1}	square millimeter per gram, mm^2 g^{-1}	0.001
	Pressure		
9.90	megapascal, MPa (10^6 Pa)	atmosphere	0.101
10	megapascal, MPa (10^6 Pa)	bar	0.1
1.00	megagram per cubic meter, Mg m^{-3}	gram per cubic centimeter, g cm^{-3}	1.00
2.09×10^{-2}	pascal, Pa	pound per square foot, lb ft^{-2}	47.9
1.45×10^{-4}	pascal, Pa	pound per square inch, lb in^{-2}	6.90×10^3

(continued on next page)

Conversion Factors for SI and non-SI Units

To convert Column 1 into Column 2, multiply by	Column 1 SI Unit	Column 2 non-SI Unit	To convert Column 2 into Column 1, multiply by
Temperature			
1.00 (K − 273)	Kelvin, K	Celsius, °C	1.00 (°C + 273)
(9/5 °C) + 32	Celsius, °C	Fahrenheit, °F	5/9 (°F − 32)
Energy, Work Quantity of Heat			
9.52×10^{-4}	joule, J	British thermal unit, Btu	1.05×10^3
0.239	joule, J	calorie, cal	4.19
10^7	joule, J	erg	10^{-7}
0.735	joule, J	foot-pound	1.36
2.387×10^{-5}	joule per square meter, J m^{-2}	calorie per square centimeter (langley)	4.19×10^4
10^5	newton, N	dyne	10^{-5}
1.43×10^{-3}	watt per square meter, W m^{-2}	calorie per square centimeter minute (irradiance), cal cm^{-2} min^{-1}	698
Transpiration and Photosynthesis			
3.60×10^{-2}	milligram per square meter second, mg m^{-2} s^{-1}	gram per square decimeter hour, g dm^{-2} h^{-1}	27.8
5.56×10^{-3}	milligram (H$_2$O) per square meter second, mg m^{-2} s^{-1}	micromole (H$_2$O) per square centimeter second, µmol cm^{-2} s^{-1}	180
10^{-4}	milligram per square meter second, mg m^{-2} s^{-1}	milligram per square centimeter second, mg cm^{-2} s^{-1}	10^4
35.97	milligram per square meter second, mg m^{-2} s^{-1}	milligram per square decimeter hour, mg dm^{-2} h^{-1}	2.78×10^{-2}
Plane Angle			
57.3	radian, rad	degrees (angle), °	1.75×10^{-2}

CONVERSION FACTORS FOR SI AND NON–SI UNITS

Electrical Conductivity, Electricity, and Magnetism

To convert Column 1 into Column 2, multiply by	Column 1 SI Unit	Column 2 non-SI Unit	To convert Column 2 into Column 1, multiply by
10	siemen per meter, S m^{-1}	millimho per centimeter, mmho cm^{-1}	0.1
10^4	tesla, T	gauss, G	10^{-4}

Water Measurement

9.73 × 10^{-3}	cubic meter, m^3	acre-inches, acre-in	102.8
9.81 × 10^{-3}	cubic meter per hour, m^3 h^{-1}	cubic feet per second, ft^3 s^{-1}	101.9
4.40	cubic meter per hour, m^3 h^{-1}	U.S. gallons per minute, gal min^{-1}	0.227
8.11	hectare-meters, ha-m	acre-feet, acre-ft	0.123
97.28	hectare-meters, ha-m	acre-inches, acre-in	1.03 × 10^{-2}
8.1 × 10^{-2}	hectare-centimeters, ha-cm	acre-feet, acre-ft	12.33

Concentrations

1	centimole per kilogram, cmol kg^{-1} (ion exchange capacity)	milliequivalents per 100 grams, meq 100 g^{-1}	1
0.1	gram per kilogram, g kg^{-1}	percent, %	10
1	milligram per kilogram, mg kg^{-1}	parts per million, ppm	1

Radioactivity

2.7 × 10^{-11}	becquerel, Bq	curie, Ci	3.7 × 10^{10}
2.7 × 10^{-2}	becquerel per kilogram, Bq kg^{-1}	picocurie per gram, pCI g^{-1}	37
100	gray, Gy (absorbed dose)	rad, rd	0.01
100	sievert, Sv (equivalent dose)	rem (roentgen equivalent man)	0.01

Plant Nutrient Conversion

	Elemental	Oxide	
2.29	P	P$_2$O$_5$	0.437
1.20	K	K$_2$O	0.830
1.39	Ca	CaO	0.715
1.66	Mg	MgO	0.602

1 Integrating Mineralizable Nitrogen Indices into Fertilizer Nitrogen Recommendations

Charles W. Rice and John L. Havlin

Kansas State University
Manhattan, KS

In most agricultural soils and cropping systems, additional fertilizer, manure, or legume N must be supplied to optimize crop yields. Before the production of nitrogenous fertilizers in the 1940s, farmers relied upon legume crops, manure, and soil organic matter as the primary N sources to support crop production. Mineralization of soil organic N historically has been a significant source of crop N; however, with increased crop productivity, mineralized N from soil usually cannot meet the entire N needs of a crop. Quantification of mineralized N from the soil has regained value for efficient use of N to lower production costs and decrease risks to the environment.

Determination of the fertilizer N requirement of crops depends on an estimate of the quantity of N needed by the crop and an estimate of both inorganic and organic soil N sources. Nitrogen rates are commonly determined by models represented by:

$$\text{N recommendation} = a(\text{yield goal}) - b(\text{soil test N}) - c(N_{min}) \qquad [1]$$

In the above model, yield goal influences the quantity of fertilizer N recommended more than any other term. Thus, determination of optimum fertilizer N rates requires good estimates of realistic yield goals for each field (Jackson et al., 1987). Yield goals usually are determined by adding 5 to 10% to the average yield during the last 5 to 7 yr. Results from North Dakota suggest that nearly 80% of growers overestimate their yield goal (Goos & Prunty, 1990). In a 4-yr Nebraska study, only 10% of 158 farmers surveyed reached their yield goal, and half produced <80% of their yield goal (Schepers et al., 1986).

Soil test N (Eq. [1]) represents extractable inorganic N and is used in making N recommendations in the Great Plains and other arid environments. This test quantifies the soil profile NO_3^- or inorganic N content at sampling time and is subject to considerable error in fields where NO_3^- is lost prior to plant uptake (Hergert, 1987).

The N mineralization (N_{min}) term in Eq. [1] should include adjustments for N_{min} from previous legume crops, manure applications, soil organic matter, and

Copyright © 1994 Soil Science Society of America, 677 S. Segoe Rd., Madison, WI 53711, USA.
Soil Testing: Prospects for Improving Nutrient Recommendations, SSSA Special Publication 40.

other organic N sources. Currently, most N recommendation models do not explicitly include estimates of N_{min}, despite the development of many laboratory and greenhouse procedures to quantify it (Keeney, 1982). The omission of N_{min} is due to our inability to accurately assess N mineralization. The actual amount of N mineralized during a growing season varies according to soil properties and weather. Unfortunately, field calibration of the N_{min} indices or field data quantifying the contribution of potentially mineralizable N to the N requirement of the crop are limited (Meisinger, 1984).

The objectives of this chapter are to: (i) provide an overview of current indices of N mineralization; (ii) summarize calibration of N_{min} indices; (iii) discuss factors that regulate N mineralization; and (iv) provide recommendations for future research and development in the use of N_{min} in N management. Subsequent chapters will focus on particular techniques of estimating N mineralization and integrating it into N fertilizer recommendations.

INDICES OF NITROGEN MINERALIZATION

The historical indices of N mineralization can be divided into laboratory, field, and modeling procedures. The goal of the laboratory procedures has been to develop a quick routine test for estimating the pool of mineralizable N that would correlate to field N mineralization. A more comprehensive review of N mineralization tests has been presented previously (Stanford, 1982; Keeney, 1982; Meisinger, 1984).

Laboratory

Chemical

Numerous chemical extractants have been evaluated as indices of N_{min} (Keeney, 1982). Extractants include strong acids or bases, neutral salts, and water, and procedures vary in extractant concentration and extraction time and temperature (Table 1-1). The analytes measured in these extractants are distillable NH_3-N, NH_4-N, and total N.

Table 1-1. Common chemical extraction procedures used to estimate N_{min}.

Extractant	Concentration	Time	Temperature
	M	h	°C
Acids/Bases			
H_2SO_4	0.25 to 6	1 to 28	25
HCl	0.1 to 5	1 to 26	25
NaOH	1	0.5 to 4.2	25 to 100
Salts			
KCl	1 to 2	1 to 20	80 to 100
$CaCl_2$	0.01	1 to 16	100 to 121
K_2SO_4	0.01 to 0.5	1	25 to 100
$NaHCO_3$	0.01 to 0.1	0.25	25
$KMnO_4$	(alkaline)	0.25	100
Water			
—	—	1	100

In general, the strong acids and bases extract larger quantities of N than do salt or water extractants, and usually are correlated highly with total N. The mild extractants remove some or all of the active soil N pool, which usually are highly correlated with N uptake under greenhouse conditions (Table 1–2). The recommended procedure uses 0.01 M CaCl$_2$ to extract NH$_4$–N after 16 h in an autoclave at 121°C (Keeney, 1982). Although chemical extractant procedures are relatively rapid and inexpensive, they have not been used routinely as estimates of N$_{min}$ because they have not been calibrated with field estimates of N$_{min}$. Development of a chemical extractant that imitates a microbiological process seems unlikely.

Biological

Biological procedures can be divided into short-term, long-term, and greenhouse techniques. Short-term methods usually include some variation of aerobic or anaerobic incubations (Table 1–3). These techniques are discussed in greater detail elsewhere (Stanford, 1982; Keeney, 1982). Longer term incubations are used to better characterize the mineralizable N pool (N$_0$) and to quantify mineralization kinetics. Variations of the long-term incubations include use of

Table 1–2. Correlations between selected measures of N$_{min}$.

Comparison			Correlation coefficient (r)
Kansas (Aiman, 1992)			
2 M KCl (100°C,4h)	vs.	aerobic incubation	0.81**
2 M KCl (100°C,4h)	vs.	N uptake (greenhouse)	0.62**
Total N	vs.	N uptake (greenhouse)	0.83**
Organic C	vs.	N uptake (greenhouse)	0.74**
Pennsylvania (Fox and Piekielek, 1984)			
Total N	vs.	anaerobic incubation	0.79
CaCl$_2$ Extr.†	vs.	anaerobic incubation	0.74
CaCl$_2$ Extr.,UV‡	vs.	N uptake (field)	0.37
CaCl$_2$ Extr.,UV§	vs.	N uptake (field)	0.49
CaCl$_2$ Extr.,UV¶	vs.	N uptake (field)	0.65
North/South Dakota (Gelderman et al., 1988)			
CaCl$_2$ Extr.†	vs.	Total N	0.73**
CaCl$_2$ Extr.	vs.	N uptake (field)	0.21

**Significant at the 0.01 probability level.
†0.01 M CaCl$_2$ (121°C, 16 h).
‡All 67 fields.
§55 fields (poorly drained soils omitted).
¶45 fields (poorly drained and previous legume fields omitted).

Table 1–3. Common biological procedures used to estimate N$_{min}$ in the laboratory.

Procedure	Temperature	Time	Measurement
	°C	wk	
Aerobic	25–35	2–3	NH$_4^+$ + NO$_3^-$
	25–35	3–2	NH$_4^+$ + NO$_3^-$
Anaerobic	30–40	1–3	NH$_4^+$

disturbed and undisturbed soil cores; however, incubating disturbed soil samples results in higher estimates of N_{min} than undisturbed samples (Cabrera & Kissel, 1988). Nitrogen mineralization determined from aerobic or anaerobic incubations are highly correlated to N uptake measured in greenhouse experiments (Stanford, 1982). Unfortunately, N_{min} determined from incubations is not well correlated to N uptake estimated from field experiments (Table 1–2; Fox & Piekielek, 1984). The recommended biological procedure is to measure NH_4^+ produced in a 7-d anaerobic incubation at 40°C.

The primary disadvantages of microbial methods are the influence of sample handling and pretreatment effects on N_{min} and the relatively long incubation times required (Meisinger, 1984). As with any chemical or biological N_{min} estimate, the value for predicting N requirements of crops is realized only after calibration with field estimates.

Greenhouse methods use measures of grain yield, aboveground dry matter, N concentration, and N uptake in the harvested material. Greenhouse techniques provide a good screening tool, but they have limited application to field conditions. As with all greenhouse methods used for assessing nutrient availability, N_{min} determined from greenhouse studies usually are poorly correlated to N uptake or availability measured in field experiments. One main disadvantage is the relatively long incubation time required for plant growth and analysis.

Field

Soil

Several field indices also have been used to quantify N mineralization. Field measures can be divided into soil incubations and plant uptake approaches. The field incubations are listed in Table 1–4. The buried bag and ion exchange resin methods have been used in native ecosystems, particularly forest ecosystems, to estimate N mineralization in situ, but have not been used extensively in agriculture. The buried bag was proposed by Eno (1960) and has since been modified (DiStefano & Gholz, 1986). Anion exchange resins or membranes can be inserted into the soil and incubated for a designated time to accumulate mineralized N (Schnabel, 1983; Qian et al., 1993). These techniques show some promise and need to be evaluated.

The most promising techniques are variations of the soil NO_3^- test. This technique measures inorganic N, usually NO_3^-, at a specific stage of crop growth, based

Table 1–4. Field indices of N mineralization.

Soil	Plant
Buried bag	whole plant sample
intact soil samples	ear leaf, flag leaf, or petiole
mixed soil samples	chlorophyll content
Ion exchange resins	stalk tests: NO_3^-
anion: NO_3^-	
cation: NH_4^+	
Soil nitrate tests	
preplant	
presidedress	

on crop N needs and time of N fertilizer application (Magdoff et al., 1984; Magdoff, 1991). Two suggested sampling times are preplant and presidedress for corn. Sampling for soil NO_3^- includes residual soil NO_3^- from the previous season, N mineralization and any losses that may have occurred through leaching or denitrification. Unfortunately the soil NO_3^- test cannot predict N mineralization after sampling. These tests are discussed further by Meisinger and Schepers (1994, Chapter 3).

Plant

The plant can be the integrator of N dynamics. Indices used are listed in Table 1–4. Sampling of whole plants or specific plant parts are used to assess N nutrition of the plant. Plant indices are difficult to incorporate into adjustment of fertilizer recommendations because of the time constraints. Sampling of leaves such as flag leaf or ear leaf, often occurs too late in the growing season to apply N fertilizer. Plant sampling usually involves only the aboveground biomass, which will underestimate the total amount of N mineralized because the roots may contain a significant amount of N. More recent techniques, such as measurement of the chlorophyll content, have shown some promise. These techniques will be discussed in greater detail in Chapter 3. Plant N uptake provides the best opportunity to integrate the soil, plant, and weather dynamics into N mineralization, but several disadvantages preclude widespread use.

Field evaluation of N_{min} by plants is the most precise method and is used as the standard by which other procedures are compared. Nitrogen uptake is usually measured on N control plots and represents inorganic and mineralized organic soil N. If inorganic N is measured at planting and after harvest, N_{min} represents the difference between total N uptake and the change in inorganic N in the soil profile. At least 2 yr of field study are needed, because residual fertilizer N may be present in the first year (Broadbent & Carlton, 1978). Measuring only N uptake in grain is not appropriate, because grain yield can be influenced strongly by environmental conditions during grain fill.

The inherent limitations on the use of plant tests for quantifying N_{min} are (i) considerable time and labor commitments, (ii) unique site specificity, and (iii) significant year by location interactions. Despite these disadvantages, field (and perhaps greenhouse) studies are essential for calibrating chemical or biological estimates of N_{min} for use in N recommendation models.

Relationships between Nitrogen Mineralization Indices

Successful use of N_{min} in predicting crop N requirements depends on calibration of an index or estimate of N_{min} measured in the field. It is well established that many chemical and biological indices are highly correlated with N uptake in greenhouse studies. For example, Aiman (1992) reported that KCl extractable NH_4–N (100°C, 4 h) was highly correlated to N_{min} determined through aerobic incubation and N uptake by wheat in the greenhouse (Table 1–2). Total N and organic C, however, were equally effective measures of N_{min}. Similar results were reported by Fox and Piekielek (1984) using anaerobic incubation and 0.01 M $CaCl_2$ (121°C, 16 h) to estimate N_{min} in 67 soils (Table 1–2).

Although a few studies have presented encouraging results (Fox & Piekielek, 1978), laboratory estimates of N_{min} generally have not been highly correlated to field measures of N_{min} (Table 1–2). In the study by Fox and Piekielek (1984), 0.01 M CaCl$_2$ (121°C, 16 h, ultraviolet) extractable NH$_4$–N was not well correlated to N uptake in corn; however, when the poorly drained soils and those fields planted to legumes within the preceding 2 yr were omitted, the correlations improved. Gelderman et al. (1988) also reported poor correlations between laboratory and field estimates of N_{min} in 69 field studies with winter wheat in North Dakota and South Dakota (Table 1–2). These data demonstrate the importance of extensive field calibrations that include characterization of soil properties and environmental conditions (i.e., moisture and temperature).

Modeling

A promising but less developed approach is modeling. Current models vary in complexity (Tanji et al., 1979; Van Veen et al., 1981; Molina et al., 1983; Broadbent, 1986). Some simple models include the use of mineralization potential (N_0) and rate (k):

$$N_{min} = N_0 (1 - e^{-kt}) \qquad [2]$$

where N_{min} = the amount of N mineralized
N_0 = the mineralizable N pool
k = the rate constant
t = time

The rate constant can be adjusted for soil water content (Stanford & Epstein, 1974) and temperature (Stanford et al., 1975). With this approach, Smith et al. (1977) reported good correlation to N mineralized in the field under fallow conditions. Others also have reported favorable results (Herlihy, 1979; Marion et al., 1981). Cabrera and Kissel (1988) used a similar approach, but reported very poor prediction of N mineralized under field conditions as measured by plant N uptake. The model overestimated N mineralization by 60 to 300% and was attributed to differences in clay and soil water contents (Cabrera & Kissel, 1988). Campbell et al. (1988) also developed a model in which the N mineralization rate constant was adjusted for temperature and moisture. Their model performed well under irrigated conditions, but underestimated N mineralization under dryland conditions. They speculated that the model did not correctly predict the effect of wetting and drying cycles on flushes of N mineralization. The inclusion of heat units also has been proposed as a possible technique to predict N mineralization (Honeycutt et al., 1988). Modeling will be discussed further in subsequent chapters.

REGULATION OF NITROGEN MINERALIZATION

Despite the intensive research on N mineralization, why are we still unable to incorporate N_{min} indices into N recommendations? Many indices fail to consider the factors that regulate N_{min} under field conditions. Nitrogen mineralization is regulated by many important factors (Fig. 1–1). The size of the organic N or mineralizable pool determines the amount of N potentially mineralized, but substrate quality, moisture, temperature, and accessibility will modify the mineralization rate.

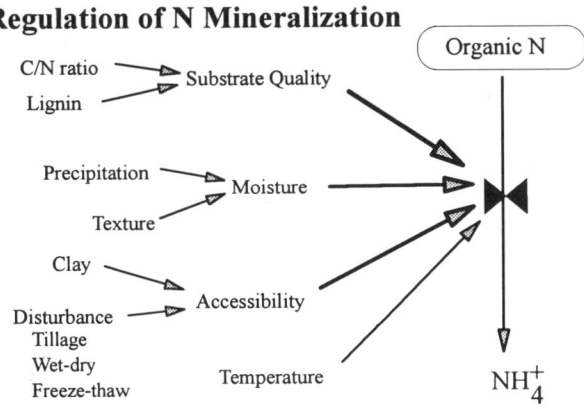

Fig. 1–1. Concept of regulatory factors that affect N mineralization.

Organic Nitrogen

The size of the mineralizable pool is important in determining the amount of N mineralized. Most studies have shown that mineralization follows first order kinetics and is dependent on the size or concentration of the organic N pool (Stanford & Smith, 1972). Subsequent research has used Eq. [2] or some modification (Broadbent, 1986; Juma et al., 1984; Campbell et al., 1988). This previous research indicates that the size of the mineralizable pool will determine the amount of mineralization.

Substrate Quality

The quality of the substrate being mineralized also affects the amount of net N mineralization. Substrate quality is characterized by the C/N ratio or the lignin content. A poor quality substrate slows mineralization and increases the competition for any organic N ammonified. Competition for NH_4^+ stimulates reassimilation, resulting in less net N mineralization. Bonde and Rosswall (1987) suggested that the mineralization rate constant is influenced by the quality of the substrate. This is supported by using a two-compartment model to represent rapid and slow mineralizable N pools. Seasonal variation in N_0 also suggests substrate differences on the N mineralization rate (El-Harris et al., 1983; Bonde & Rosswall, 1987).

Soil Water Content and Aeration

Soil water content is a key factor regulating N mineralization. Aerobic microbial activity is optimal at ≈60% water-filled pore space (Linn & Doran, 1984). Maximum N mineralization and immobilization should occur at this value. As soil water content increases, aeration becomes limiting, resulting in anaerobic activity. Nitrogen mineralization and immobilization still occur, but anaerobic rates are slower than those under aerobic conditions. Because of lower N requirements by microbes under anaerobic conditions, NH_4^+ usually accumulates. Optimum soil moisture for N mineralization is between –0.01 and –0.03 MPa

(Myers et al., 1982; Doel et al., 1990) Within the range of field capacity (−0.03 MPa) to wilting point (−1.5 MPa), Stanford and Epstein (1974) reported that N mineralization was related linearly to percentage of soil water content. Myers et al. (1982) confirmed the linear relationship between −0.03 and −4.0 MPa for most soils, although a curvilinear response was measured for a few soils. Moisture variations may account for significant variation in N mineralization (Campbell et al., 1974; Kowanlenko & Cameron, 1976; Cassman & Munns, 1980).

Accessibility

Accessibility is the availability of the organic N to the microorganisms. It is well known that clay protects organic matter from mineralization. Tillage also will change the accessibility of the organic N to the microbes. Nitrogen mineralization is commonly stimulated after tillage by the breaking up of soil aggregates and exposing previously protected organic N. Tillage effects are apparent with no-tillage and reduced tillage systems that exhibit lower mineralization rates (Dowdell & Cannell, 1975; Powlson, 1980; Rice et al., 1987; Sarrantonio & Scott, 1988). Aggregate disruption under laboratory conditions increases net N mineralization (Craswell & Waring, 1972; Cabrera & Kissel, 1988). Cycles of wetting-drying and freezing-thawing also will stimulate N mineralization. The stimulation is due to disruption of soil aggregates and partial killing of the microbial community (Seneviratne & Wild, 1985; Sparling & Ross, 1988; West et al., 1992). Rewetting of dried soils generally stimulates microbial activity (West et al., 1992), which can increase N mineralization. Van Gestel et al. (1993) attributed the N mineralization flush partly to killed biomass during drying and partly to nonbiomass organic residues. The size of the N flush was influenced by organic C and clay contents of soil. Cabrera (1993) suggested that the flush of N mineralization during rewetting increased the rapidly mineralizable pool and that N models could be modified by increasing the pool size.

Temperature

Because N mineralization is a biological process, temperature regulates it. Temperature effects on N mineralization are based on the change in the reaction rate for every 10° increase in temperature, $Q_{10} = 2$ for the temperature range of 0 to 35°C (Stanford et al., 1975; Campbell et al., 1981). Other studies, however, have shown a different relationship. Campbell et al. (1984) reported that Q_{10} varied across climatic zones ranging from 2.00 to 2.75, with sandy soils having a higher Q_{10} than loams. Campbell et al. (1984) suggested that this difference may reflect the differences in quality of the soil organic matter more than climatic adaption by the soil microbial community. These temperature relationships have been determined in the laboratory at constant temperatures. In the field, fluctuating soil temperatures may affect N mineralization. Because the temperature response is not linear, N mineralization may vary under a fluctuating temperature regime (Campbell & Biederbeck, 1972; Myers, 1975). The response to fluctuating temperatures is greater for nitrification than for ammonification. Because ammonification is the rate-limiting step, N mineralization probably will not be affected greatly by fluctuating soil temperatures (Myers, 1975; Stanford et al., 1975).

Current Mineralization Indices

How well do mineralization indices incorporate the factors that affect N mineralization? Laboratory procedures generally fail to include environmental factors that regulates N mineralization rate. Chemical extracts characterize only the size of the mineralizable N pool and do not measure any of the factors that regulate the rate. Biological laboratory procedures attempt to measure the mineralizable N pool and may give some evidence of substrate quality by developing a rate constant. Most of the biological incubations, however, are conducted at optimum soil water and temperature conditions and will not include effects of these factors on the N mineralization rate. These laboratory procedures usually do not quantify the accessibility factor.

Field approaches have the potential of including the effects of regulatory factors on the mineralization rate. Soil incubations in the field are better able to integrate environmental factors affecting N mineralization including substrate quality, temperature, and soil water to some degree. The soil may have higher soil water contents under fallow conditions than under growing crops, and this will tend to overestimate N mineralization. Buried bag techniques include the soil water content at the time the soil was placed in the bag, but the soil water content will not fluctuate to the extent observed under field conditions. The significance of the difference in soil water content will be a function of the frequency of measurement and the local weather. Using the plant as an integrator offers the best opportunity to integrate all of the factors regulating N mineralization, but has the disadvantage of the time needed for plant N uptake and analysis to modify N fertilizer recommendations.

We believe modeling offers the best opportunity to assess N mineralization. Current models, however, do not include several of the factors that regulate the rate of N mineralization. Soil temperature and water content are sometimes used to adjust the mineralization rate. Differences in substrate quality usually are not incorporated directly into the models unless they include crop residues. Models with a crop residue component generally use the C/N ratio as an indicator of substrate quality. Accessibility of the organic N currently is not incorporated into models and probably will be the most difficult. The paucity of research on accessibility will prevent rapid development of models. Cabrera (1993) did attempt to model the flush of N mineralization after rewetting, but his solution was to enlarge the N pool following first order kinetics. This approach does not provide the mechanisms or include other factors that influence the size of the flush, such as clay content and extent of drying. Research is needed on effect of tillage, freeze–thaw, and wet–dry cycles on N mineralization.

FUTURE DEVELOPMENT

Field indices and modeling will provide the best opportunities to integrate N mineralization into fertilizer recommendations. Research efforts should be directed to calibration of field indices by using crop N uptake. Extensive laboratory and field research is needed to develop the parameters for the N models. Mechanistic models are needed to understand the N cycle and environmental factors that influence transformations. General models are needed for producers, extension agents, and agricultural advisors to develop N fertilizer recommendations. Models would make it possible to calculate crop growth and N uptake and quantitatively include the various sources of N.

A conceptual model of this approach is illustrated in Fig. 1–2. Soil N models that are climate driven (temperature and moisture) could incorporate residual N as well as N mineralization potential to predict profile NO_3^- content at planting time. The fertilizer N recommendation would be determined subsequently by a current model that includes yield goal and N availability from legumes, manure, or organic matter. An accurate recommendation of N fertilizer rate needs to be based on a reliable estimate of crop N needs. Initially, long-term weather data could be used, for example, to predict N fertilizer needs in the fall and enhance N manufacturing, distribution, and marketing efficiency. With an interactive model, final adjustments in the N rate could be made again in the spring by using real-time over winter weather data, before actual N fertilizer application. In this example, a grower would receive an initial N recommendation in the fall based on long-term over winter climatic data and, subsequently, an updated N recommendation based on actual overwinter climatic conditions. Only with continual developmental and applied research in quantifying and predicting N_{min} under field conditions, however, can models like the one suggested be developed. This model should include any NO_3^- lost through leaching and denitrification during the nongrowing season. The amount of N mineralized during the nongrowing season could be confirmed by soil N tests or, once confidence is developed, the model could be used to predict residual soil inorganic N. This type of modeling approach has been used with some success for potatoes in the Netherlands (Neeteson et al., 1989). The deficiency of modeling and other indices is the weather during the late growing season whose effects on N dynamics and plant N needs cannot be predicted.

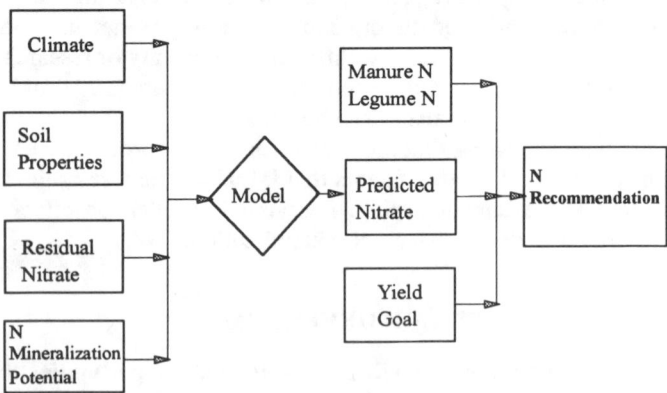

Fig. 1–2. Conceptual model of how N mineralization in model could be integrated into N fertilizer recommendations. Climate data could be the long-term, 30-yr average or actual overwinter climate.

ACKNOWLEDGEMENTS

This chapter was supported partly by a specific cooperative agreement from the USDA-ARS no. 58-5440-2-105. Contribution no. 94-326-B from the Kansas Agric. Exp. Stn., Manhattan, KS.

REFERENCES

Aiman, N. 1992. Evaluation of the chemical methods for estimating potential mineralizable organic nitrogen in soils. M.S. thesis. Kansas State Univ., Manhattan.

Bonde, T.A., and T. Rosswall. 1987. Seasonal variation of potentially mineralizable nitrogen in four cropping systems. Soil Sci. Soc. Am. J. 51:1508–1514.

Broadbent, F.E. 1986. Empirical modeling of soil nitrogen mineralization. Soil Sci. 141:208–213.

Broadbent, F.E., and A.B. Carlton. 1978. Field trails with isotopically labeled nitrogen fertilizer. p. 1–41. *In* D.R. Nielsen and J.G. MacDonald (ed.) Nitrogen in the environment. Vol. 1. Nitrogen behavior in field soil. Academic Press, New York.

Cabrera, M.L. 1993. Modeling the flush of nitrogen mineralization caused by drying and rewetting soils. Soil Sci. Soc. Am. J. 57:63–66.

Cabrera, M.L., and D.E. Kissel. 1988. Potentially mineralizable nitrogen in disturbed and undisturbed soil samples. Soil Sci. Soc. Am. J. 52:1010–1015.

Campbell, C.A., and V.O. Biederbeck. 1972. Influence of fluctuating temperatures and constant soil moistures on nitrogen changes in amended and unamended loam. Can. J. Soil. Sci. 52:332–336.

Campbell, C.A., Y.W. Jame, and R. De Jong. 1988. Predicting net nitrogen mineralization over a growing season: Model verification. Can. J. Soil Sci. 68:537–552.

Campbell, C.A., Y.W. Jame, and G.E. Winkleman. 1984. Mineralization rate constants and their use for estimating nitrogen mineralization in some Canadian prairie soils. Can. J. Soil Sci. 64:333–343.

Campbell, C.A., R.J.K. Myers, and K.L. Weir. 1981. Potentially mineralizable nitrogen, decomposition rates and their relationship to temperature for five Queensland soils. Aust. J. Soil Res. 19:323–332.

Campbell, C.A., D.W. Stewart, W. Nicholaichuk, and V.O. Biederbeck. 1974. Effects of growing season soil temperature, moisture, and NH_4-N on soil nitrogen. Can. J. Soil Sci. 54:403–412.

Cassman, K.G., and D.N. Munns. 1980. Nitrogen mineralization as affected by soil moisture, temperature and depth. Soil Sci. Soc. Am. J. 44:1233–1236.

Craswell, E.T., and S.A. Waring. 1972. Effect of grinding on the decomposition of soil organic matter: I. The mineralization of organic nitrogen in relation to soil type. Soil Biol. Biochem. 4:427–433.

DiStefano, J.F., and H.L. Gholz. 1986. A proposed use of ion exchange resins to measure nitrogen mineralization and nitrification in intact soil cores. Commun. Soil Sci. Plant Anal. 17:989–998.

Doel, D.S., C.W. Honeycutt, and W.A. Halterman. 1990. Soil water effects on the use of heat units to predict crop residue carbon and nitrogen mineralization. Biol. Fert. Soils 10:102–106.

Dowdell, R.J., and R.Q. Cannell. 1975. Effect of ploughing and direct drilling on soil nitrate content. J. Soil Sci. 26:53–61.

El-Harris, M.K., V.L. Cochran, L.F. Elliot, and D.F. Bezdicek. 1983. Effect of tillage, cropping and fertilizer management on soil nitrogen mineralization potential. Soil Sci. Soc. Am. J. 47:1157–1161.

Eno, C.F. 1960. Nitrate production in the field by incubating the soil in polyethylene bags. Soil Sci. Soc. Am. Proc. 24:277–279.

Fox, R.H., and W.P. Piekielek. 1978. Field testing of several nitrogen availability indexes. Soil Sci. Soc. Am. J. 42:747–750.

Fox, R.H., and W.P. Piekielek. 1984. Relationships among anaerobically mineralized nitrogen, chemical indexes, and nitrogen availability to corn. Soil Sci. Soc. Am. J. 48:1087–1090.

Gelderman, R.H., W.C. Dahnke, and L. Swenson. 1988. Correlation of several N indices for wheat. Commun. Soil Sci. Plant Anal. 19:755–772.

Goos, R.J., and L. Prunty. 1990. Yield variability and the yield goal decision. p. 187–189. *In* J.L. Havlin (ed.) Proc. of the Great Plain Soil Fertility Conf., Denver, CO. 6–7 Mar. 1990. Kansas State Univ., Manhattan.

Hergert, G.W. 1987. Status of residual nitrate-nitrogen soil tests in the United States of America. p. 73–89. *In* J.R. Brown (ed.) Soil testing: Sampling, correlation, calibration, and interpretation. SSSA Spec. Publ. 21. SSSA, Madison, WI.

Herlihy, M. 1979. Nitrogen mineralization in soils of varying texture, moisture and organic matter: I. Potential and experimental values in fallow soils. Plant Soil 53:225–267.

Honeycutt, C.W., L.M. Zibilske, and W.M. Clapham. 1988. Heat units for describing carbon mineralization and predicting net nitrogen mineralization. Soil Sci. Soc. Am. J. 52:1346–1350.

Jackson, G., D. Keeney, D. Curwen, and B. Webendorfer. 1987. Agricultural management practices to minimize groundwater contamination. Environ. Resources Center, Dep. of Nat. Resour., Univ. of Wisconsin, Madison, WI.

Juma, N.G., E.A. Paul, and B. Mary. 1984. Kinetic analysis of net nitrogen mineralization in soil. Soil Sci. Soc. Am. J. 48:753–757.

Keeney, D.R. 1982. Nitrogen availability indices. p. 711–733. *In* A.L. Page et al. (ed.) Methods of soil analysis. Part 2. 2nd ed. Agron. Monogr. 9. ASA and SSSA, Madison, WI.

Kowanlenko, C.G., and D.R. Cameron. 1976. Nitrogen transformation in an incubated soil as affected by combinations of moisture content and temperature and adsorption-fixation of ammonium. Can. J. Soil. Sci. 56:63–70.

Linn, D.M., and J.W. Doran. 1984. Effect of water filled pore space on CO_2 and N_2O production in tilled and nontilled soils. Soil Sci. Soc. Am. J. 48:1267–1272.

Magdoff, F. 1991. Understanding the Magdoff pre-sidedress nitrate test for corn. J. Prod. Agric. 4:297–305.

Magdoff, F.R., D. Ross, and J. Amadon. 1984. A soil test for nitrogen availability to corn. Soil Sci. Soc. Am. J. 48:1301–1304.

Marion, G.M., J. Kummerow, and P.C. Miller. 1981. Predicting nitrogen mineralization in chaparral soils. Soil Sci. Soc. Am. J. 45:956–961.

Meisinger, J.J. 1984. Evaluating plant-available nitrogen in soil-crop systems. p. 391–416. *In* R.D. Hauck (ed.) Nitrogen in crop production. ASA, CSSA, and SSSA, Madison, WI.

Molina, J.A.E., C.E. Clapp, M.J. Schaffer, F.W. Chichester, and W.E. Larson. 1983. NCSOIL, a model of nitrogen and carbon transformations in soil: Description, calibration, and behavior. Soil Sci. Soc. Am. J. 47:85–91.

Myers, R.J.K. 1975. Temperature effects on ammonification and nitrification in a tropical soil. Soil Biol. Biochem. 7:83–86.

Myers, R.J.K., C.A. Campbell, and K.L. Weir. 1982. Quantitative relationship between net nitrogen mineralization and moisture content of soils. Can. J. Soil Sci. 62:111–124.

Neeteson, J.J., K. Dilz, and G. Wijen. 1989. N-fertilizer recommendations for arable crops. p. 253–263. *In* J.C. Germon (ed.) Management systems to reduce impact of nitrates. Elsevier Applied Science, New York.

Qian, P., J.J. Schoenau, and A. Braul. 1993. Assessing mineralizable soil organic N and S using anion exchange. p. 257. *In* Agronomy abstracts. ASA, Madison, WI.

Powlson, D.S. 1980. Effect of cultivation on the mineralization of nitrogen in soil. Plant Soil 57:137–142.

Rice, C.W., J.H. Grove, and M.S. Smith. 1987. Estimating soil net nitrogen mineralization as affected by tillage and soil drainage due to topographic position. Can. J. Soil Sci. 67:513–520.

Sarrantonio, M., and T.W. Scott. 1988. Tillage effects on nitrogen availability to corn following a winter green manure crop. Soil Sci. Soc. Am. J. 52:1661–1668.

Schepers, J.S., K.D. Frank, and C. Bourg. 1986. Effect of yield goal and residual soil nitrogen concentrations on N fertilizer recommendations for irrigated maize in Nebraska. J. Fert. Issues. 3:133–139.

Schepers, J.S., and J.J. Meisinger. 1994. Field indicators of nitrogen mineralization. p. 31–46. *In* J.L. Havlin et al. (ed.) Soil testing: Prospects for improving nutrient recommendations. SSSA Spec. Publ. 40. SSSA, Madison, WI.

Schnabel, R.R. 1983. Measuring nitrogen leaching with ion exchange resin: A laboratory assessment. Soil Sci. Soc. Am. J. 47:1041–1042.

Seneviratne, R., and A. Wild. 1985. Effect of mild drying on the mineralization of soil nitrogen. Plant Soil 84:175–179.

Smith, S.J., L.B. Young, and G.E. Miller. 1977. Evaluation of soil nitrogen mineralization potentials under modified field conditions. Soil Sci. Soc. Am. J. 41:74–76.

Sparling, G.P., and D.J. Ross. 1988. Microbial contributions to the increased nitrogen mineralization after air drying of soils. Plant Soil 105:163–167.

Stanford, G. 1982. Assessment of soil nitrogen availability. p. 651–688. *In* F.J. Stevenson (ed.) Nitrogen in agricultural soils. Agron. Monogr. 23. ASA, CSSA, and SSSA, Madison, WI.

Stanford, G., and E. Epstein. 1974. Nitrogen mineralization water relations in soils. Soil Sci. Soc. Am. Proc. 38:103–106.

Stanford, G.A., M.H. Frere, and R.A. Vander Pohl. 1975. Effect of fluctuating temperatures on soil nitrogen mineralization. Soil Sci. 119:222–226.

Stanford, G., and S.J. Smith. 1972. Nitrogen mineralization potential in soils. Soil Sci. Soc. Am. Proc. 36:465–472.

Tanji, K.K., F.E. Broadbent, M. Mehran, and M. Fried, 1979. An extended version of a conceptual model for evaluating annual nitrogen leaching losses from croplands. J. Environ. Qual. 8:114–120.

Van Gestel, M., R. Merckx, and K. Vlassak. 1993. Microbial biomass responses to soil drying and rewetting: The fate of fast- and slow-growing microorganisms in soils from different climates. Soil Biol. Biochem. 25:109–123.

Van Veen, J.A., W.B. McGill, H.W. Hunt, M.J. Frissel, and C.V. Cole. 1981. Terrestrial nitrogen cycles: Processes, ecosystem strategies and management impacts. Ecol. Bull. 33:25–48.

West, A.W., G.P. Sparling, C.W. Feltham, and J. Reynolds. 1992. Microbial activity and survival in soils dried at different rates. Aust. J. Soil Res. 30:209–222.

2 Potential Nitrogen Mineralization: Laboratory and Field Evaluation

M.L. Cabrera and D.E. Kissel
University of Georgia
Athens, Georgia

M.F. Vigil
Central Great Plains Research Station
USDA-ARS,
Akron, Colorado

Nitrogen fertilizer rates are currently estimated without considering field to field variability in N mineralization. The use of average N rates in fields with different N mineralization often leads to inadequate as well as excessive N applications. Therefore, the capability to adjust N fertilizer rates according to N mineralization potential is required for efficient N use and for prevention of groundwater pollution with NO_3.

The adjustment of N rates for N mineralization can be made with the budget or balance-sheet method (Neeteson, 1990):

$$N_{rec} = Y_{exp} \times b - N_{up} - N_{ini} - N_{add} - M_s - M_r - M_m + N_{im} + N_{lch} + N_{gas} + N_{inf} \quad [1]$$

where N_{rec} is the recommended N rate, Y_{exp} is the expected grain yield, b is the total N uptake per unit of grain yield (including vegetative parts and roots), N_{up} is the amount of N already absorbed by the crop at the time of fertilizer application, N_{ini} is the initial amount of inorganic soil N (at the time of fertilizer application), N_{add} represents N additions through rain and N fixation, M_s, M_r, and M_m are the amounts of net N mineralized (between the time of fertilizer application and the time at which net N uptake is complete) from soil organic matter, residue, and manure, respectively, N_{im} is fertilizer N immobilized, N_{lch} and N_{gas} are leaching and gaseous N losses, and N_{inf} is the amount of inorganic N at the end of the season. Thus, knowledge of several variables is needed for calculating correct N rates; however, the only variables that can be known accurately at the time of fertilizer application are inorganic N (N_{in}) and N accumulated by the crop up to that time (N_{up}). All other variables, including N mineralized from soil organic matter (M_s) and from crop residues (M_r), must be estimated.

Copyright © 1994 Soil Science Society of America, 677 S. Segoe Rd., Madison, WI 53711, USA.
Soil Testing: Prospects for Improving Nutrient Recommendations, SSSA Special Publication 40.

The adjustment of N rates for N mineralization can also be made with simulation models that take into account plant growth, nutrient uptake, and soil N transformations (Osmond et al., 1992). Simulation models also use an N balance approach, but in addition require a quantitative understanding of the biological processes involved. Consequently, the modeling approach can provide a finer resolution of the fertilizer N required for economically and environmentally sound crop production. Operationally, to calculate optimum fertilizer rates with simulation models, the user would need to simulate crop yield at several N fertilizer rates, fit the resulting yields to an equation of the form: Yield = f(N applied) and then compute the economic optimum N rate in the traditional manner (Heady, 1956).

While the balance-sheet method requires estimates of N mineralized from soil organic matter, crop residues, and manures, simulation models require estimates of the sizes and rates of N mineralization of those N pools (Willigen & Neeteson, 1985; Wolf et al., 1989). In this chapter, we review research on estimating sizes and rates of N mineralization of soil organic matter pools, as well as amounts of N mineralized from soil organic matter and crop residues.

NITROGEN MINERALIZATION FROM SOIL ORGANIC MATTER

Magnitude

The amount of N mineralized from soil organic matter depends on soil type and environmental conditions. Some estimates of net N mineralized during the corn (*Zea mays* L.) growing season range from 13 to 131 kg N ha^{-1} in Nebraska (Saint-Fort et al., 1990) and from 50 to 123 kg N ha^{-1} in Japan (Saito & Ishii, 1987). Field studies conducted for 2 yr in Kansas (Cabrera & Kissel, 1988a) showed that N mineralized during the sorghum [*Sorghum bicolor* (L.) Moench] growing season varied from 31 to 51 kg N ha^{-1} in three soils. In the same study, N mineralized during ~3 mo in two fallowed soils ranged from 75 to 107 kg N ha^{-1} due to the higher water content of fallowed than cropped soils. Campbell et al. (1988) reported that N mineralized from a Canadian soil during 92 d was 52, 81, and 86 kg N ha^{-1} under dryland cropped, irrigated cropped, and summer fallow treatments, respectively. These results show a large variability in N mineralization and indicate that in many cases N mineralized from soil organic matter is a significant proportion of the amount of N needed by a crop.

Estimation of Potentially Mineralizable Nitrogen with Laboratory Incubations

Potentially mineralizable N (N_0) refers to an active fraction of soil N that is mainly responsible for the release of inorganic N through microbial activity. Stanford and Smith (1972) proposed a method for estimating N_0 that involves incubating a soil–sand mixture under optimum temperature and moisture conditions. The soil–sand mixture is leached periodically (0.01 M CaCl$_2$) to remove inorganic N, and cumulative N mineralized with time (excluding the first 2 wk) is fitted to a single exponential model of the form: $N_{min} = N_0(1 - e^{-kt})$, where N_{min} is cumulative N mineralized in time t, N_0 is an initial estimate of potentially mineralizable N, and k is its rate constant of mineralization. The final value for N_0 is calculated by adding the amount of N mineralized during the first 2 wk to the value of N_0 estimated with the model. Stanford and Smith (1972) used this procedure to estimate N_0 because the single exponential model could not adequately describe the initial flush of N mineralization observed in many soils. Later, other workers suggested modifying the single exponential model to include a constant amount of N that mineralizes immediately after starting incubation (Jones, 1984), or after a fixed period of time (Beauchamp et al., 1986). These models provided a better fit to mineralization data that show an initial flush.

In an attempt to find better models to describe N mineralized, Molina et al. (1980) and Deans et al. (1986) used two first-order equations (double exponential model) to represent two pools of mineralizable N, one of them decomposing much faster than the other and accounting for the initial flush of N mineralization observed in many soils. Deans et al (1986) showed that this model was better than the single exponential model for describing data obtained in several laboratory studies.

Several authors (Juma et al., 1984; Ellert & Bettany, 1988; Bonde & Lindberg, 1988; White & Marinakis, 1991) reviewed other regression models for describing N mineralization, some of which can adequately describe the lag that occurs when the mineralization rate depends not only on substrate concentration, but also on the size of the microbial population. Houot et al. (1989) also proposed quantifying N_0 by adjusting a simulation model to reproduce experimental results and using as N_0 the initial N content of some of the organic N pools. Nevertheless, the single and double exponential models are the most used ones.

Several factors affect the size and rate constant of mineralization of pools identified with the single and double exponential models of N mineralization. Among them are the regression procedure used, length of incubation, date of soil sampling, and soil depth.

Effect of Regression Procedure

Stanford and Smith (1972) estimated N_0 and k using an iterative procedure to fit their data to a logarithmic transformation of the single exponential model. Their fitting method has the drawback of weighting the largest values of cumulative N mineralized more than the smallest numbers. Smith et al. (1980) compared the procedure used by Stanford and Smith (1972) to a nonlinear regression technique and concluded that when the errors are uncorrelated with sampling time, the nonlinear technique had lower root mean square values than the linear method of Stanford and Smith (1972). Similarly, Talpaz et al. (1981) compared linear regression after log transformation to a nonlinear regression procedure for estimating N_0 and k for the 39 soils used by Stanford and Smith (1972). They found that for all soils, N_0 was smaller, k was greater, and the sum of squares of the error was smaller with the nonlinear technique than with the linear procedure.

Ellert and Bettany (1988) showed that fitting a nonlinear model to incremental data from each incubation interval is better than fitting the model to cumulative data obtained by summing incremental observations. Some of the advantages of using an incremental model are that the interdependence of observation errors is decreased and that it is possible to use data with missing observations. Thus, the use of nonlinear regression procedures, such as PROC NLIN in SAS

(SAS Institute, 1985), to fit incremental models may provide the most simple and precise approach to estimate parameter values for the single and double exponential models of N mineralization.

Effect of Length of Incubation Time

Paustian and Bonde (1987) observed that mineralization curves with different values of N_0 and k can appear very similar for short incubations, and noted that parameters N_0 and k could change as incubation time increased. Similarly, Cabrera and Kissel (1988b) observed that as incubation time increased (from 140 to 252 d), the size of the mineralizable N pools estimated with the double exponential model increased whereas the corresponding rate constants decreased. These results emphasize (i) the importance of standardizing incubation times so that comparison of pool sizes and rates can be made without the confounding effect of time, and (ii) the need to consider both N_0 and k as indicators of the mineralizing capacity of soils.

Effect of Date of Sample Collection

Few studies have evaluated the effect of date of sampling on N_0 and k. El-Haris et al. (1983) determined N_0 and k in spring and fall samples taken from a long-term crop rotation-tillage experiment. They found that the average N_0 was 72 mg N kg^{-1} in the spring and 131 mg N kg^{-1} in the fall, whereas the respective average k were 0.349 and 0.096 wk^{-1}. Bonde and Rosswall (1987) measured N mineralized in samples taken from four cropping systems in April, June, August, and October. They observed that the N mineralized in a 13-wk incubation decreased from April through August and increased from August to October. The increase observed after August was attributed to the addition of crop residues, which probably enhanced N immobilization and increased the amount of N in dead cells and microbial metabolites. Although there is no direct evidence indicating that these forms of N may constitute a significant proportion of mineralizable N in soils, a close association was found between mineralizable N and microbial biomass (Bonde et al., 1988).

In a recent unpublished study, we measured N_0 (by the method of Gianello & Bremner, 1986a), microbial biomass, inorganic N, and water content in soil samples collected monthly (for 18 mo) from the upper layer (0–2.5, 2.5–5, 5–10, and 10–15 cm) of plots under conventional and no-till management on a Cecil sandy clay loam (clayey, kaolinitic, thermic Typic Kanhapludult) in Georgia. Seasonal changes in N_0 were very evident in the 0 to 2.5 cm of no-till plots, and were highly correlated with changes in microbial biomass N and C (Fig. 2–1). These seasonal changes appeared to follow changes in soil water content. In contrast to the results obtained in soil from no-till plots, soil from conventional-till plots did not show defined seasonal changes in potentially mineralizable N and microbial biomass.

In summary, because N_0 may change during the year, it is important to collect soil samples at the beginning of the season for which N_0 is to be estimated. It is clear that more research is needed to understand the factors that affect these changes.

Effect of Soil Depth

Because organic N in soils decreases with depth, it is expected that mineralizable N would also decrease with depth. Stanford et al. (1974) found that potentially mineralizable N, expressed both in absolute amount and as a percentage of total N, decreased with depth in the upper three 15-cm layers of eight Idaho soils.

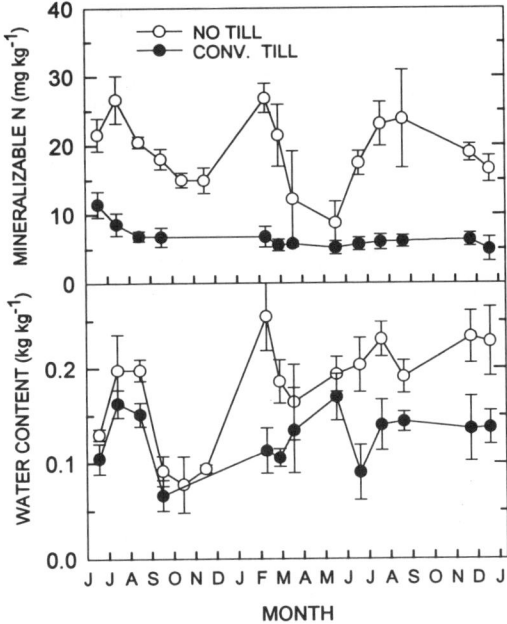

Fig. 2-1. Evolution of potentially mineralizable N and soil water content in the upper 2.5 cm of no-till and conventional-till plots on a Cecil sandy clay loam in Georgia.

Hadas et al. (1986) also found that potentially mineralizable N decreased with depth in several Israeli soils. They did not find any consistent effect of depth on the rate constant of mineralization. Cabrera (1986) fitted the double-exponential model to data from the four upper 15-cm layers (0- –60-cm depth) of five Kansas soils, and observed that the slow pool of mineralizable N (expressed as % of total N) decreased with depth, whereas the fast pool remained constant or tended to increase with depth. If the fast pool of mineralizable N is assumed to be caused by drying the soil before the incubation, then the observed increase with depth in the percentage of the total soil N constituting the fast pool may be explained by fewer drying–rewetting cycles and less intense drying in the lower soil layers than in the surface ones. Soils with repeated drying–rewetting cycles show a decrease in the magnitude of the flushes due to a depletion of the organic reserves (Birch, 1958). Therefore, soils with fewer drying–rewetting cycles would be expected to show relatively larger flushes of N mineralization than soils with repeated cycles of drying and rewetting.

Hadas et al. (1986) estimated that the relative contribution of the top 20 cm of soil to the total N mineralization in the upper 60 cm varied from 45 to 75%. Similarly, in a 13-wk incubation, Cassman and Munns (1980) found that the total N mineralized from the upper 18-cm layer made up 42% of the total N mineralized in the upper 108 cm of soil. Cabrera (1986) estimated that the top 15 cm of soil contributed from 45 to 56% of the total N mineralized in 120-cm profiles of three Kansas soils, and that ~90% was mineralized in the top 45 cm.

The results summarized above indicate that subsurface layers may significantly contribute to the total N mineralized in a soil and may need to be considered when estimating N mineralized during a growing season. The relative importance of N mineralized from subsurface layers may increase when moisture is limiting in the surface soil, but not in subsurface layers.

Estimation of Potentially Mineralizable Nitrogen and Rate Constant from Chemical and Physical Tests

Stanford and Smith (1972) and Oyanedel and Rodríguez (1977) concluded that a common k value could be used for all the soils they studied. According to their results, only N_0 would need to be estimated to predict N mineralized in the field. This prompted a search for chemical methods of determining N_0 that would not be as time-consuming as long-term incubations. Stanford (1982) summarized studies that showed that N_0 could be predicted from the NH_4–N extracted by autoclaving the soil for 16 h in 0.01 M $CaCl_2$ (Stanford & Smith, 1976), or from the NH_4–N released by treating the soil with acid permanganate (Stanford & Smith, 1978). Recently, Gianello and Bremner (1986b) developed two chemical methods of soil N availability that showed high correlation with N_0. One method consists of determining the NH_3–N produced by steam distilling the soil sample with phosphate–borate buffer solution of pH 11.2. The other method involves measuring the NH_4–N produced during treatment of the soil with 2 M KCl at 100°C for 4 h. Also, a high correlation between total N and N_0 was found by some workers (Marion et al., 1981; Gianello & Bremner, 1986b).

Recent results indicate that k may not remain constant within a soil profile or among soils (Griffin & Laine, 1983; Juma et al., 1984; Hadas et al., 1986; Cabrera & Kissel, 1988a). These results suggest the need for rapid methods to estimate both N_0 and k. Cabrera and Kissel (1988c) found that N_0 and k for 16 undisturbed soil samples could be estimated from total N and clay contents. Similarly, Myers (1989) reported that N_0 and k values for 30 undisturbed soil cores could be predicted from cation-exchange capacity and total N and clay contents. Clearly, more work is needed to elucidate the nature of organic pools contained in N_0 so that better predictive equations and better methods of measurement can be developed.

Use of Nitrogen Mineralization Potentials and Rate Constants to Predict Nitrogen Mineralized in the Field

Values for N_0 and k are developed from laboratory incubations at optimum temperature and soil water content. To use these parameters for estimating N mineralization in the field, it is necessary to apply correction factors for field temperature and soil water content.

Stanford et al. (1973a) found that the effect of temperature on k could be described with a modified Arrhenius equation. A pooled regression for 39 soils ($\log_{10} k = 6.16 - 2299/T$) indicated Q_{10} values (change in reaction rate for every 10°C increase in temperature), of 1.9, 1.8, and 1.8 for 5 to 15, 15 to 25, and 25 to 35°C, respectively. Campbell et al. (1981) also used a modified Arrhenius equation to describe the effect of temperature on k in five Australian soils. They found that for samples from the upper 15 cm of soil, the average Q_{10} values were 1.88, 1.81, and 1.74 for 10 to 20, 20 to 30, and 30 to 40°C, respectively ($\log_{10} k = 6.14 - 2285/T$). For samples from the 15- to 30-cm layer, Q_{10} values were 1.68, 1.62, and 1.57 for the same temperature ranges described above ($\log_{10} k = 4.81 - 1861/T$).

Addiscott (1983) studied the effect of temperature on k in three British soils. He used a zero-order kinetics model and found that the Arrhenius equation properly described the dependence of k on temperature. The Q_{10} values measured between 5 and 15°C ranged from 2.31 to 3.18, whereas those observed between 15 and 25°C varied between 1.81 and 2.55.

In a subsequent study, Campbell et al. (1984) determined Arrhenius relationships between k and absolute temperature for 33 Canadian surface (0–15 cm) soils from five different zones. Because there were no significant differences in Arrhenius relationships among soils within each soil zone, average equations were developed for each zone (average Q_{10} values ranged from 2.0 to 2.75). Comparing their results with those of other workers, Campbell et al. (1984) noted that Q_{10} values tend to increase from warm climates to colder climates, possibly an indication of the resistance to degradation of the organic materials in soils of different climates. Thus, determination of Q_{10} values for a given area may be necessary to accurately represent the effect of temperature on N mineralization.

Prediction of N mineralized in the field also has to take into account the effect of soil water content. Stanford and Epstein (1974) used nine U.S. soils to study the relationship between soil water content and N mineralization. They found that if the values of cumulative N mineralized were expressed on a relative basis with respect to maximum N mineralized, then the effect of water content could be described as:

Relative N mineralized = soil water content/optimum water content

Similar results were obtained by Cavalli and Rodríguez (1975) with nine Chilean soils. In contrast, Myers et al. (1982) noted that the above expression was not suitable for five Australian and 32 Canadian soils. Instead, they proposed a model of the following form, constrained to pass through $X = 0$, $Y = 0$; $X = 1$, $Y = 1$:

$$Y = bX + (1 - b) X^2$$

where Y   = relative N mineralized
       X   = [SWC – SWC(–4.0 MPA)]/[OWC – SWC(–4.0 MPa)]
       b   = regression coefficient
       SWC   = soil water content
       SWC(–4.0 MPa) = SWC at –4.0 MPa
       OWC   = optimum water content (about –0.03 MPa)

They found that in one Australian and nine Canadian soils, N mineralized showed a curvilinear response to X, a relationship that was described with b values ranging from 1.22 to 2.18. For the other soils b was equal to 1, giving a linear response with respect to X.

Therefore, it would appear that in certain cases the simple approach proposed by Stanford and Epstein (1974) may not be adequate to represent the effect of soil water content. In addition, Cassman and Munns (1980) showed that N mineralization may be affected by a soil water content × temperature interaction. This interaction is not considered in the model proposed by Stanford and Smith (1972).

There is some debate concerning the method used to correct for soil water content. In some studies (Campbell et al., 1988; Griffin & Laine, 1983), the water content factor, y, has been used to correct k as in: $N_{min} = N_0 (1 - e^{-kyt})$, but Olness (1984) suggested using y to correct N mineralized as in: $N_{min} = N_0 (1 - e^{-kt}) y$. Campbell et al. (1988) and Griffin and Laine (1983) observed that the first model produced slightly larger predictions than the second one.

The correction for temperature and soil water content has been performed using hourly (Campbell et al., 1984, 1988), daily (Cabrera & Kissel, 1988a), weekly (Smith et al., 1977; Marion et al., 1981), monthly (Oyanedel & Rodríguez, 1977; Stanford et al., 1977; Prado & Rodríguez, 1978), and yearly (Verstraete & Voets, 1976) time steps. Also, most workers have used Euler's integration method, whereas Campbell et al. (1984, 1988) used the integrated form of the model to estimate N mineralized within each time step.

Stanford (1982) summarized the first studies that attempted predictions of N mineralized from soil organic matter under greenhouse (Stanford et al., 1973b) and under field conditions (Prado & Rodríguez, 1978; Smith et al., 1977). We shall concentrate on subsequent studies.

Marion et al. (1981) used an approach similar to that proposed by Stanford and Smith (1972) to predict annual N mineralized in two chaparral soils. They incubated soil samples from the upper 30 cm of soil (in 10-cm increments) and determined N_0 and k by fitting the experimental data to a model of the form:

$$\log_{10}(N_0 - N_{min}) = \log_{10}N_0 - kt^b,$$

where b is a regression coefficient and all other variables and parameters are as defined earlier. Predicted N mineralized agreed to within ±10% with independent estimates of N mineralized based on N balance equations.

Campbell et al. (1984) used N_0 and k values for the top 2.5 cm of a loam soil to predict N mineralized in the laboratory and in the field, taking into account soil water and temperature conditions. The model predicted well the N mineralized in the laboratory and tended to underpredict the N mineralized in the field at the two lowest water contents. These underpredictions were attributed to the accidental entry of water into some bags, which could have caused wetting and drying cycles and induced flushes of N mineralization (Campbell et al., 1988).

In another study, Campbell et al. (1988) conducted a more rigorous field evaluation of Stanford and Smith's (1972) approach. They used N_0 and k values for the upper two 7.5-cm layers of a loam soil, together with estimated temperature and soil water content, to predict N mineralized in lysimeters (15 cm diam., 120 cm long) subjected to summer fallow, dryland wheat (*Triticum aestivum*, L) and irrigated wheat treatments. Predicted N mineralized agreed with measured values during the first 45 to 60 d, and tended to underpredict observed values afterwards. Underpredictions for the entire growing season amounted to 16, 26, and 31% for irrigated wheat, summerfallow, and dryland wheat, respectively. The authors explained the underpredictions as due to the model not accounting for the flushes in N mineralization that typically occur as a result of wetting-drying cycles.

In a Japanese study, Saito and Ishii (1987) measured N mineralized from the organic matter of four soils during the corn growing season. They estimated N_0 and k values and evaluated the dependence of k on temperature to predict N mineralized, adjusted for field temperature. In most soils, predicted amounts of N mineralized agreed closely with observed values in 1983, and tended to overpredict measured values in 1984 (Table 2–1). The authors attributed part of the overpredictions to the existence of dry conditions in 1984. Because the model used did not have adjustments for soil water content, it would be expected to overpredict N mineralized under dry field conditions.

Cabrera and Kissel (1988a) evaluated Stanford and Smith's (1972) approach in three Kansas soils during two sorghum seasons. Using the upper three 15-cm layers (0- –45-cm total depth), they obtained overpredictions ranging from 114 to 343% of the actual amounts of N mineralized (32–51 kg N ha^{-1}). When only the upper 15-cm layer was considered, the overpredictions ranged from 7 to 181% (Table 2–2; Cabrera, 1986). Cabrera and Kissel (1988a) hypothesized that the overpredictions could have been due to an improper function for adjustment of the amounts of N mineralized based on soil water content, and to the crushing of the soil samples (to pass 2-mm sieve) before incubation. In a follow-up study, Cabrera and Kissel (1988c) showed that even after subtracting the initial N flush, samples that had been dried and crushed to pass through a 2-mm sieve showed higher N mineralization than undisturbed soil samples. These results suggest that with some soils it may be necessary to use undisturbed, field moist samples for determination of mineralization parameters.

In summary, the limited research results available indicate that it may be possible to predict N mineralized under field conditions following Stanford and Smith's (1972) approach. It would appear, however, that best results can be obtained if the soils samples are disturbed as little as possible to avoid an artificial enhancement of N mineralization. In addition, it may be necessary to include

Table 2–1. Measured and predicted N mineralized from four Japanese soils in two corn growing seasons (131–142 d) (adapted from Saito & Ishii, 1987).

Year	N Mineralized (kg ha^{-1})		% Error
	Measured	Predicted[†]	
	Ishidoriya		
1983	80	74	–8
1984	115	145	26
	Kuriyagawa		
1983	86	85	–1
1984	113	127	12
	Matsukawa		
1983	50	52	4
1984	93	131	41
	Toshima		
1983	83	103	24
1984	123	124	1

[†] Predicted considering only the upper 20 cm of each soil.

Table 2–2. Measured and predicted N mineralized from three Kansas soils in two sorghum growing seasons (105–162 d) (adapted from Cabrera, 1986).

Year	N Mineralized (kg ha^{-1})		% Error
	Measured	Predicted†	
	Haynie		
1983	43	58	35
1984	46	49	7
	Kahola		
1983	32	90	181
1984	51	107	110
	Richfield		
1983	31	69	123
1984	36	50	39

†Predicted considering only the upper 15 cm of each soil.

the effects of drying and rewetting cycles to simulate N mineralization in soils where these cycles are common. Cabrera (1993) showed that in addition to causing an initial flush of N mineralization, drying and rewetting soils may also increase background mineralization when compared with soils that remain moist. More research in this area is needed to understand the mechanisms involved.

NITROGEN MINERALIZATION FROM CROP RESIDUES

Several complete reviews summarized how soil water content, temperature, and the physical and chemical nature of a crop residue affect the amount and rate of N mineralized or immobilized when that residue is mixed with soil (Bartholomew, 1965; Smith & Peterson, 1982; Sommers & Giordano, 1984). How we relate this information to fertilizer recommendations is still open to question. In the previous section, we described two methods for estimating N fertilizer rates, the N balance method, and the use of a simulation model. Both methods require estimates of N mineralization from crop residues. The next section will summarize some of the recent literature concerned with modeling and predicting the amounts and rates of crop residue N mineralization.

Magnitude

The amount of N mineralized or immobilized from crop residues will depend on soil type (Christensen, 1985, 1986), residue characteristics (Smith & Peckenpaugh, 1986; Vigil & Kissel, 1991), and environmental conditions (Boddy, 1983). Field experiments showed that residues with C/N ratios >50 initially immobilize N and eventually release a relatively low amount of N to subsequent crops (White et al., 1986; Seligman et al., 1986). In contrast, residues with C/N ratios <25 can supply considerable amounts of N to subsequent crops (Bowen et al., 1993; Wagger, 1989; Ranells & Wagger, 1992).

Estimating Nitrogen Mineralized from Crop Residues

Vigil and Kissel (1991) evaluated plant residue C, N, and lignin for the prediction of total seasonal net N mineralization. From that work, several linear regression models were developed to estimate the maximum amount of N that may mineralize in a season from incorporated crop residues of different N contents. The best fitted model was of the form: $Y = 0.62 + 1.338\,N - 0.875\,L/N$, where Y is the amount of N mineralized expressed as a percentage of the residue N applied, and L and N are lignin and N concentrations (g kg^{-1}) in the residue, respectively. Whereas this equation predicts the seasonal net N mineralization, it is not valuable for assessing short-term mineralization or mineralization kinetics. Soil environmental factors such as soil water content, temperature, and crop residue management would all alter the rate of N mineralization. The use of a simulation model to predict N mineralization rates makes it possible to take into account soil environment and soil or crop residue management factors.

Vigil et al. (1991) evaluated a simulation model to predict field N mineralization of ^{15}N-labeled sorghum residues incorporated into a Smolan silt loam (fine, montmorillonitic, mesic Pachic Argiustoll) in Kansas. In that study, a N transformation subroutine derived from Version 2.0 of the CERES-Maize model (Jones & Kiniry, 1986) closely predicted the N mineralized after 110 d of decomposition (Fig. 2–2). Bowen et al. (1993) further evaluated the N submodel from CERES-Maize version 2.10 and found agreement between measured and simulated values of N mineralized from legume green manures incorporated into Brazilian soils.

In the two studies mentioned above, the model was tested for accuracy in simulating N mineralized from residues incorporated into the soil. In no-till and minimum-till systems, however, crop residues are either left intact on the soil surface, or are only partially buried. Additional modeling studies are needed to assess decomposition and N mineralization under those conditions. This requires the development of methods to measure or predict soil and crop residue water contents and temperature at the soil surface because these variables are major factors controlling microbial activity.

Bristow et al. (1986) developed a model for simulating heat and water transfer through a surface residue-soil system. Stroo et al. (1989) used Bristow's

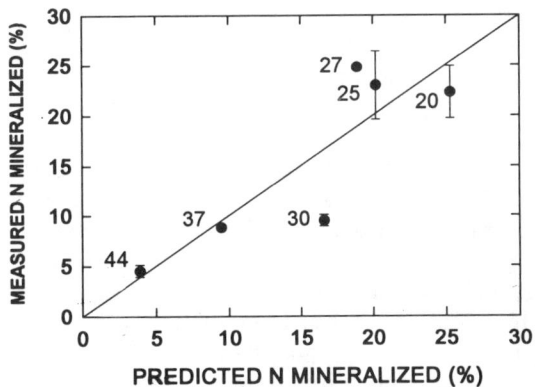

Fig. 2–2. Measured and predicted (by CERES N submodel) N mineralized from sorghum residues after 110 d of field decomposition in a Smolan silt loam in Kansas (numbers next to symbols indicate the C/N ratio of the sorghum residue; adapted from Vigil et al., 1991).

model and the model of Knapp et al. (1983) to develop a model for predicting residue decomposition at the soil surface. Stroo's model accurately predicted the mass loss of surface-placed wheat residues at both Bushland, TX, and Lafayette, IN. While the model calculates N mineralized or immobilized, it has not been independently validated for its ability to accurately assess N mineralization and immobilization of surface-placed residues.

The results reviewed above indicate that simulation models are promising tools for predicting N mineralized from crop residues. Further research and development of models is needed to improve their predicting capability.

CONCLUSIONS

Nitrogen mineralization is a complex process that depends on soil properties, crop residue quantity, and quality, as well as climatic conditions. Because of this complexity, computer simulation models will be very useful tools in the future for gaining a better understanding of the rate of N mineralization in the field. We also believe that these models can be used in soil testing programs for direct estimation of the amount of N that will be mineralized in individual farmers' fields.

The best methodology to model N mineralization from soil organic matter at this time is with long-term, multi-week incubations that allow one to determine the potentially mineralizable N for a soil, N_0, as well as its rate constant of mineralization, k. Such a procedure, however, is not a practical method for our traditional soil testing programs that require rapid determinations and completed reports on soil nutrient availability within 2 or 3 d. It appears more likely that the model parameters needed for computer simulation of soil N mineralization could be determined from a quick soil test. Research is needed to evaluate the feasibility of this approach.

In order to make progress on N mineralization from crop residues, we must continue to refine the appropriate models. We cannot directly test a crop residue for N mineralization, so we must continue to study how crop residue characteristics such as its composition of lignin or hemicellulose affect the rate of decomposition and the N mineralization rate. Because of the recent emphasis on reduced tillage and no-till, we need to understand better how crop residues placed on the soil surface will behave with regard to rates of decomposition and N mineralization. The contact between residue and soil is clearly less under no-till management and the array of decomposing organisms will be different than for the case of crop residue buried in the soil. We must also understand and predict the temperature and water content more accurately at the soil surface so that we can provide the correct values of temperature and water potential for driving these models.

Finally, since N fertilizer recommendations are typically made prior to planting the crop, and since there is considerable uncertainty in future weather conditions (and therefore uncertainty in predicting future soil temperature and soil water content), it will not be possible to predict with high accuracy the amount of N that will be mineralized. It will be possible, however, to quickly simulate the estimated N mineralized for several years, having a range of climatic conditions. This would allow one to determine an average estimate of the

N likely to be mineralized. The other possibility is to use real time climatic data for early in the growing season to predict the N mineralized up to the time when N fertilizer is applied in a sidedress treatment. A similar modeling approach might also have considerable value in cropping systems that have fallow periods (such as in the Great Plains) during which N can mineralize and accumulate in the soil. The climatic conditions during these fallow periods and the soil and crop residue conditions can vary considerably, depending on the location and the year, and therefore result in quite different amounts of N mineralized. By using records of the climate in real time and by having measurements or estimates of crop residue properties, reasonably accurate simulations of N mineralizations should be possible under fallow conditions.

In summary, we believe that efforts should continue to gain a better understanding of the complex process of N mineralization, which supplies a substantial, and in some cases most of the N needed by crops. Our ability to predict the quantity of N mineralized in soils has been much improved by recent research. Further efforts to improve our ability to predict N mineralized under field conditions will help to increase the efficiency of use of N fertilizers to achieve economically and environmentally sound crop production.

REFERENCES

Addiscott, T.M. 1983. Kinetics and temperature relationships of mineralization and nitrification in Rothamsted soils with differing histories. J. Soil Sci. 34:343–353.

Bartholomew, W.V. 1965. Mineralization and immobilization of nitrogen in the decomposition of plant and animal residues. p. 285–306. *In* W.V. Bartholomew and F.E. Clark (ed.) Soil nitrogen. Agron. Monogr. 10. ASA, CSSA, and SSSA, Madison, WI.

Beauchamp, E.G., W.D. Reynolds, D. Brasche-Villeneuve, and K. Kirby. 1986. Nitrogen mineralization kinetics with different soil pretreatment and management histories. Soil Sci. Soc. Am. J. 50:1478–1483.

Birch, H.F. 1958. The effect of soil drying on humus decomposition and nitrogen availability. Plant Soil 10:9–31.

Boddy, L. 1983. Carbon dioxide release from decomposing wood: Effect of water content and temperature. Soil Biol. Biochem. 15:501–510.

Bonde, T.A., and T. Lindberg. 1988. Nitrogen mineralization kinetics in soil during long-term aerobic laboratory incubations: A case study. J. Environ. Qual. 17:414–417.

Bonde, T.A., and T. Rosswall. 1987. Seasonal variation of potentially mineralizable nitrogen in four cropping systems. Soil Sci. Soc. Am. J. 51:1508–1514.

Bonde, T.A., J. Schnurer, and T. Rosswall. 1988. Microbial biomass as a fraction of potentially mineralizable nitrogen in soils from long-term field experiments. Soil Biol. Biochem. 20:447–452.

Bowen, W.T., J.W. Jones, R.J. Carsky, and J.O. Quintana. 1993. Evaluation of the nitrogen submodel of CERES-Maize following legume green manure incorporation. Agron. J. 85:153–159.

Bristow, K.L., G.S. Campbell, R.I. Papendick, and L.F. Elliot. 1986. Simulation of heat and moisture transfer through a surface residue-soil system. Agric. Forest Meteorol. 36:193–214.

Cabrera, M.L. 1986. Studies on the prediction of nitrogen mineralized from soil organic matter under field conditions. Ph.D. diss. Kansas State Univ. (Diss. Abstr. DA8705829).

Cabrera, M.L. 1993. Modeling the flush of nitrogen mineralization caused by drying and rewetting soils. Soil Sci. Soc. Am. J. 57:63–66.

Cabrera, M.L., and D.E. Kissel. 1988a. Evaluation of a method to predict nitrogen mineralized from soil organic matter under field conditions. Soil Sci. Soc. Am. J. 52:1027–1031.

Cabrera, M.L., and D.E. Kissel. 1988b. Length of incubation time affects the parameter values of the double exponential model of nitrogen mineralization. Soil Sci. Soc. Am. J. 52:1186–1187.

Cabrera, M.L., and D.E. Kissel. 1988c. Potentially mineralizable nitrogen in disturbed and undisturbed soil samples. Soil Sci. Soc. Am. J.52:1010–1015.

Campbell, C.A., Y.W. Jame, and R. de Jong. 1988. Predicting net nitrogen mineralization over a growing season: Model verification. Can. J. Soil Sci. 68:537–552.

Campbell, C.A., Y.W. Jame, and G.E. Winkleman. 1984. Mineralization rate constants and their use for estimating nitrogen mineralization in some Canadian prairie soils. Can. J. Soil Sci. 64:333–343.

Campbell, C.A., R.J.K. Myers, and K.L. Weier. 1981. Potentially mineralizable nitrogen, decomposition rates and their relationship to temperature for five Queensland soils. Aust. J. Res. 19:323–332.

Cassman, K.G., and D.N. Munns. 1980. Nitrogen mineralization as affected by soil moisture, temperature and depth. Soil Sci. Soc. Am. J. 44:1233–1237.

Cavalli, J.C., and J. Rodríguez. 1975. Effect of moisture on nitrogen mineralization of nine soils of Santiago Province. Cienc. Invest. Agrar. 2:101–111.

Christensen, B.T. 1985. Wheat and barley straw decomposition under field conditions: Effect of soil type and plant cover on weight loss, nitrogen and potassium content. Soil Biol. Biochem. 17:691–697.

Christensen, B.T. 1986. Barley straw decomposition under field conditions: Effect of placement and initial nitrogen content on weight loss and nitrogen dynamics. Soil Biol. Biochem. 18:523–529.

Deans, J.R., J.A.E. Molina, and C.E. Clapp. 1986. Models for predicting potentially mineralizable nitrogen and decomposition rate constants. Soil Sci. Soc. Am. J. 50:323–326.

El-Haris, M.K., V.L. Cochran, L.F. Elliot, and D.F. Bezdicek. 1983. Effect of tillage, cropping, and fertilizer management on soil nitrogen mineralization potential. Soil Sci. Soc. Am. J. 47:1157–1161.

Ellert, B.H., and J.R. Bettany. 1988. Comparison of kinetic models for describing net sulfur and nitrogen mineralization. Soil Sci. Soc. Am. J. 52:1692–1702.

Gianello, C., and J.M. Bremner. 1986a. A simple chemical method of assessing potentially available organic nitrogen in soil. Commun. Soil Sci. Plant Anal. 17:195–214.

Gianello, C., and J.M. Bremner. 1986b. Comparison of chemical methods of assessing potentially available organic nitrogen in soil. Commun. Soil Sci. Plant Anal. 17:215–236.

Griffin, G.F., and A.F. Laine. 1983. Nitrogen mineralization in soils previously amended with organic wastes. Agron. J. 75:124–129.

Hadas, A., S. Feigenbaum, A. Feigin, and R. Portnoy. 1986. Nitrogen mineralization in profiles of differently managed soil types. Soil Sci. Soc. Am. J. 50:314–319.

Heady, E.O. 1956. Methodological problems in fertilizer use. p. 3–21. *In* E.L. Baum et al. (ed.). Methodological procedures in the economic analysis of fertilizer data. Iowa State College Press, Ames, IA.

Houot, S., J.A.E. Molina, R. Chaussod, and C.E. Clapp. 1989. Simulation by NCSOIL of net mineralization in soils from the Deherain and 36 Parcelles fields at Grignon. Soil Sci. Soc. Am. J. 53:451–455.

Jones, C.A. 1984. Estimation of an active fraction of soil nitrogen. Commun. Soil Sci. Plant Anal. 15:23–32.

Jones, C.A., and J.R. Kiniry. 1986. CERES-Maize—a simulation model of maize growth and development. Texas A&M Univ. Press, College Station.

Juma, N.G., E.A. Paul, and B. Mary. 1984. Kinetic analysis of net nitrogen mineralization in soil. Soil Sci. Soc. Am. J. 48:753–757.

Knapp, E.B., L.F. Elliot, and G.S. Campbell. 1983. Carbon, nitrogen, and microbial biomass interrelationships during the decomposition of wheat straw: A mechanistic simulation model. Soil Biol. Biochem. 15:455–461.

Marion, G.M., J. Kummerow, and P.C. Miller. 1981. Predicting nitrogen mineralization in chaparral soils. Soil Sci. Soc. Am. J. 45:956–961.

Molina, J.A.E., C.E. Clapp, and W.E. Larson. 1980. Potentially mineralizable nitrogen in soil: The simple exponential model does not apply for the first 12 weeks of incubation. Soil Sci. Soc. Am. J. 44:442–443.

Myers, R.G. 1989. Estimation of nitrogen mineralization from soil organic matter. M.S. thesis–Kansas State Univ. of Agric. and Applied Science, Manhattan.

Myers, R.J.K., C.A. Campbell, and K.L. Weier. 1982. Quantitative relationship between net nitrogen mineralization and moisture content of soils. Can. J. Soil Sci. 62:111–124.

Neeteson, J.J. 1990. Development of nitrogen fertilizer recommendations for arable crops in the Netherlands in relation to nitrate leaching. Fert. Res. 26:291–298.

Olness, A. 1984. Re: Nitrogen mineralization potentials, N_0 and correlations with maize response. Agron. J. 76:171–172.

Osmond, D.L., D.J. Lathwell, and S.J. Riha. 1992. Prediction of long-term fertilizer requirements of maize in the tropics using a nitrogen balance model. Plant Soil 143:61–70.

Oyanedel, C., and J. Rodriguez. 1977. Estimation of N mineralization in soils. Cienc. Invest. Agrar. 4:33–44.

Paustian, K., and T.A. Bonde. 1987. Interpreting incubation data on nitrogen mineralization from soil organic matter. INTECOL Bull. 15:101–112.

Prado, O., and J. Rodriguez. 1978. Estimation of nitrogen fertilizer requirements of wheat. Cienc. Invest. Agrar. 5:29–40.

Ranells, N.N., and M.G. Wagger. 1992. Nitrogen release from crimson clover in relation to plant growth stage and composition. Agron. J. 84:424–430.

Saito, M., and K. Ishii. 1987. Estimation of soil nitrogen mineralization in corn-grown fields based on mineralization parameters. Soil Sci. Plant Nutr. 33:555–566.

Saint-Fort, R., K.D. Frank, and J.S. Schepers. 1990. Role of nitrogen mineralization in fertilizer recommendations. Commun. Soil Sci. Plant Anal. 21:13–16.

SAS Institute. 1985. SAS user's guide. Statistics. Version 5 ed. SAS Inst., Cary, NC.

Seligman, N.G., S. Feigenbaum, D. Feinerman, and R.W. Benjamin. 1986. Uptake of nitrogen from high C-to-N ratio, ^{15}N-labeled organic residues by spring wheat grown under semi-arid conditions. Soil Biol. Biochem. 18:303–307.

Smith, J.H., and R.E. Peckenpaugh. 1986. Straw decomposition in irrigated soil: Comparison of twenty-three cereal straws. Soil Sci. Soc. Am. J. 50:928–932.

Smith, J.H., and J.R. Peterson. 1982. Recycling nitrogen through land application of agricultural, food processing and municipal wastes. p. 791–797. *In* F.J. Stevenson (ed.) Nitrogen in agricultural soils. Agron. Monogr. 22. ASA, CSSA, and SSSA, Madison, WI.

Smith, J.L., R.R. Schnabel, B.L. McNeal, and G.S. Campbell. 1980. Potential errors in the first-order model for estimating soil nitrogen mineralization potential. Soil Sci. Soc. Am. J. 44:996–1000.

Smith, S.J., L.B. Young, and G.E. Miller. 1977. Evaluation of soil nitrogen mineralization potentials under modified field conditions. Soil Sci. Soc. Am. J. 41:74–76.

Sommers, L.E., and P.M. Giordano. 1984. Use of nitrogen from agricultural, industrial, and municipal wastes. p. 208–220. *In* R.D. Hauck (ed.) Nitrogen in crop production. ASA, CSSA, and SSSA, Madison, WI.

Stanford, G. 1982. Assessment of soil N availability. p. 651–688. *In* F.J. Stevenson (ed.) Nitrogen in agricultural soils. Agron. Monogr. 22. ASA, CSSA, and SSSA, Madison, WI.

Stanford, G., J.N. Carter, and S.J. Smith. 1974. Estimates of potentially mineralizable soil nitrogen based on short-term incubations. Soil Sci. Soc. Am. Proc. 38:99–102.

Stanford, G., J.N. Carter, D.T. Westermann, and J.J. Meisinger. 1977. Residual nitrate and mineralizable soil nitrogen in relation to nitrogen uptake by irrigated sugarbeets. Agron. J. 69:303–308.

Stanford, G., and E. Epstein. 1974. Nitrogen mineralization-water relation in soils. Soil Sci. Soc. Am. Proc. 38:103–107.

Stanford, G., M.H. Frere, and D.H. Schwaninger. 1973a. Temperature coefficient of soil nitrogen mineralization. Soil Sci. 115:321–323.

Stanford, G., J.O. Legg, and S.J. Smith. 1973b. Soil nitrogen availability evaluations based on N mineralization potentials of soils and uptake of labeled and unlabeled nitrogen by plants. Plant Soil 39:113–124.

Stanford, G., and S.J. Smith. 1972. Nitrogen mineralization potentials of soils. Soil Sci. Soc. Am. Proc. 36:465–472.

Stanford, G., and S.J. Smith. 1976. Estimating potentially mineralizable soil nitrogen from a chemical index of soil nitrogen availability. Soil Sci. 122:71–76.

Stanford, G., and S.J. Smith. 1978. Oxidative release of potentially mineralizable soil nitrogen by acid permanganate extraction. Soil Sci. 126:210–218.

Stroo, H.F., K.L. Bristow, L.F. Elliot, R.I. Papendick, and G.S. Campbell. 1989. Predicting rates of wheat residue decomposition. Soil Sci. Soc. Am. J. 53:91–99.

Talpaz, H., P. Fine, and B. Bar-Yosef. 1981. On the estimation of N mineralization parameters from incubation experiments. Soil Sci. Soc. Am. J. 45:993–996.

Verstraete, W., and J.P. Voets. 1976. Nitrogen mineralization tests and potentials in relation to soil management. Pedologie 26:15–26.

Vigil, M.F., and D.E. Kissel. 1991. Equations for estimating the amount of nitrogen mineralized from crop residues. Soil Sci. Soc. Am. J. 55:757–761.

Vigil, M.F., D.E. Kissel, and S.J. Smith. 1991. Field crop recovery and modeling of nitrogen mineralized from labeled sorghum residues. Soil Sci. Soc. Am. J. 55:1031–1037.

Wagger, M.G. 1989. Time of dessication effects on plant composition and subsequent nitrogen release from several winter annual cover crops. Agron. J. 81:236–241.

White, G.S., and Y.D. Marinakis. 1991. A flexible model for quantitative comparisons of nitrogen mineralization patterns. Biol. Fertil. Soils 11:239–244.

White, P.J., I. Vallis, and P.G. Saffigna. 1986. The effect of stubble management on the availability of ^{15}N-labelled residual fertilizer nitrogen and crop stubble nitrogen in an irrigated black earth. Aust. J. Exp. Agric. 26:99–106.

Willigen, de P., and J.J. Neeteson. 1985. Comparison of six simulation models for the nitrogen cycle in the soil. Fert. Res. 8:157–171.

Wolf, J., C.T. de Wit, and H. van Keulen. 1989. Modeling long-term crop response to fertilizer and soil nitrogen. Plant Soil 120:11–22.

3 Field Indicators of Nitrogen Mineralization

J. S. Schepers
USDA-ARS
Lincoln, Nebraska

J. J. Meisinger
USDA-ARS
Beltsville, Maryland

Nitrogen mineralization is one of the most important soil N cycle processes in nature. Along with N immobilization, it is at the very heart of the energy producing decomposition process that drives the microbial *engine* in soils. Nitrogen mineralization is a complex process that involves a vast collection of microorganisms (bacteria, fungi, and actinomyces) acting on a wide array of substrates (crop residues, soil humus, dead microbial tissue, and manure) under varying soil environments (temperature, water content, and aeration) to produce a remarkably simple product (NO_3–N) that can be used by plants, lost to the atmosphere as N gases, immobilized, accumulated in soil, or leached from the soil-crop system. Soil scientists have studied N mineralization for decades because it was widely recognized that understanding this fundamental process is crucial to the design of efficient N management strategies in order to achieve sustained crop productivity with minimal environmental impact.

The purposes of this chapter are to: (i) review some of the principles involved in studying N mineralization, (ii) provide an overview of selected field N mineralization techniques, and (iii) delineate the situations and circumstances where these techniques can provide useful information in designing more efficient N management strategies for crop production and environmental quality.

PRINCIPLES

Nitrogen mineralization is defined as the conversion of organic N to inorganic N as a result of microbial activity (Soil Science Society of America, 1987). Nitrogen immobilization is the corollary to mineralization and is defined as the conversion of inorganic N to the organic N form in microbial tissues (Soil Science Society of America, 1987). These deceptively simple definitions conceal a very complex underlying process.

Copyright © 1994 Soil Science Society of America, 677 S. Segoe Rd., Madison, WI 53711, USA.
Soil Testing: Prospects for Improving Nutrient Recommendations, SSSA Special Publication 40.

Mineralization transforms complex organic molecules from the N pools of: soil organic matter, crop residues, microbial tissue, crop roots, manures, and other organic debris into NH_4–N and NO_3–N. The inorganic N may accumulate in the soil (net mineralization) or be converted back into microbial biomass (net immobilization) that runs the soil mineralization-immobilization turnover cycle. Which process dominates depends on the chemical nature of the compound being decomposed; compounds rich in N (C/N of less than 15:1) favor mineralization, while compounds with low N contents (C/N of greater than 25:1) favor immobilization. Virtually all field mineralization techniques measure net mineralization over time, i.e., the end result of gross mineralization vs. gross immobilization over shorter periods of time. It is well to remind ourselves, that in nature these opposing processes are inseparable. Interpretation and understanding of several of the field N mineralization techniques will hinge on recognizing this fundamental and inseparable linkage between mineralization and immobilization.

Since N mineralization is microbially driven, it is also affected by soil temperature, moisture, aeration, pH, and other physical and chemical factors. The dynamic nature of these factors in the field has led some soil scientists to study N mineralization under the controlled temperature and moisture conditions of the laboratory (e.g., Bremner, 1965; Stanford & Smith, 1972). Extrapolating these laboratory results back to field conditions, however, has met with only limited success (Stanford et al., 1977; Smith et al., 1977; Griffin & Laine, 1983). The environmental factors influencing mineralization also interact with each other (e.g., aeration × soil moisture), with soil management practices (i.e., tillage or residue placement), and with soil properties (i.e., texture or drainage) to produce the final soil microbial environment specific to a given situation. Field mineralization techniques, therefore, have the advantage of including all of these environmental interactions; but they also suffer the disadvantage of being quite site-specific. The following discussion will point out circumstances where careful interpretation is needed to properly apply and understand specific field mineralization techniques.

Another point to highlight about field mineralization principles is that NO_3–N is mobile and is subject to several fates in the soil-crop system. The major pathways for NO_3–N are crop uptake, leaching, immobilization, and denitrification. Crop uptake can effectively accumulate mineralized N under N stressed conditions—indeed this is one of the time-honored methods to estimate mineralization. Leaching and denitrification can dramatically reduce the NO_3–N accumulated from mineralization. Losses through leaching or denitrification, however, are difficult to take into account with field mineralization techniques because these processes occur as sporadic episodes associated with large water additions. Therefore, one must be cautious about interpreting N mineralization data for time intervals that include large rain or irrigation events.

The final component to highlight about field mineralization is the slow rate and the variability of the process. A high mineralizing soil, with no recent organic amendments, might release 100 kg N ha^{-1} during a typical 180 d crop season (April–September), an average of ~ 0.55 kg N ha^{-1} d^{-1}. This scenario would be typical for mineralization in the surface 30 cm of a soil containing 2% organic matter, assuming that ~3% of the organic N mineralizes annually and that about two-thirds of that amount mineralizes during the growing season. The slowness of mineralization, coupled with the normal variability of soil NO_3-N measurements (typical coefficient of variations (CVs) of 50%) combine to limit the precision and accuracy of field N mineralization estimates. Spatial variability and rate of the field mineralization processes will also be examined in the following discussion, since an intensive sampling strategy over an extended period of time will be required to obtain highly accurate field N mineralization estimates.

APPROACHES FOR QUANTITATIVE ESTIMATES

Nonfertilized Cropped Plots

Nitrogen uptake by a recently unfertilized crop remains the best and most practical estimate of field N mineralization, provided a few precautions are observed. This method has the advantage of being conducted under field temperature, moisture and aeration conditions, and it incorporates the root interactions of the specific crop species of interest (i.e., root depth, root density, or exudates). It can also be easily adapted to include effects such as: tillage practices, soil drainage, and residue management. The most difficult problem is to determine an estimate of root N, if one is interested in an absolute mineralization value. In many cases, researchers assume (knowingly or unknowingly) equal quantities of root N in fertilized and nonfertilized plots. This assumption is open to question and should be evaluated on a case-by-case basis. Ideally, root N accumulation should be directly measured, but previously reported literature values are often sufficient. Implications of assuming similar quantities of root N in fertilized and nonfertilized plots are that N deficient crops may distribute proportionately more photosynthate to the roots and therefore underestimate mineralization if the estimate is based solely on the N content of aboveground biomass.

Crop growth on check plots is usually slower than on N fertilized plots because of differences in N availability. Some growth differences can also be associated with how well net N mineralization is synchronized with crop N needs. For crops like corn (*Zea mays* L.), limited N availability early in the growing season can disproportionately reduce the grain yield potential of the crop. Nitrogen contained in corn stover (excluding roots) of adequately fertilized corn is usually ~40% of total crop N uptake, but stover N from check plots may represent a higher proportion of total N uptake. Therefore, it is important to quantify total N uptake from biomass for check plots rather than estimate N uptake based on grain yield.

The most important precaution to take with this method is to minimize N inputs from nonmineralization sources and to accurately document these other N inputs. Several cases have been reported where residual mineral N has given falsely high mineralization estimates the first year of a study (e.g., Broadbent & Carlton, 1979). First year mineralization estimates are probably the most realistic in terms of N mineralized from the previous years crop residue, but are confounded by year-to-year differences in residual N. This point is demonstrated using irrigated corn yields from first-year check plots over several years (Fig. 3–1). Total N uptake rather then grain yield would provide a better estimate of

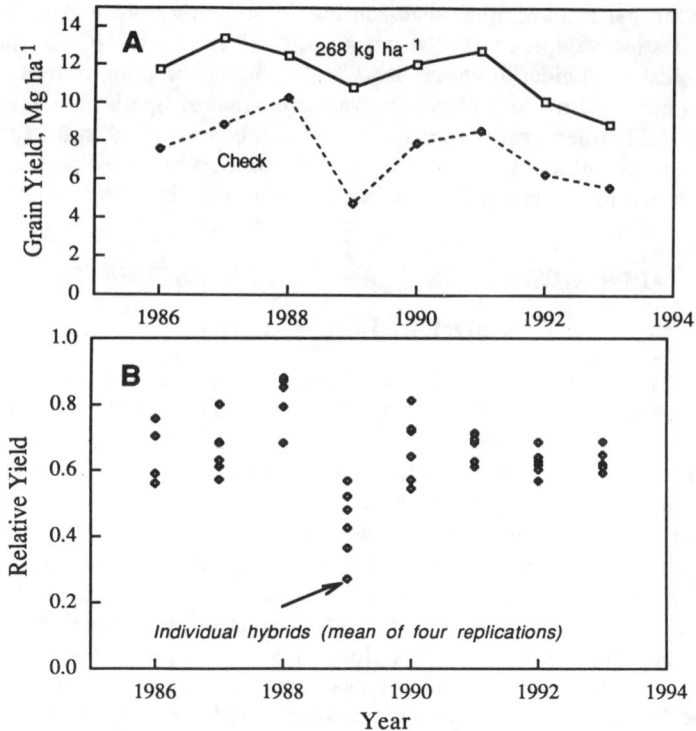

Fig. 3–1. Annual differences in irrigated corn grain yields at Schuyler, Nebraska for (A) first-year check plots and adequately fertilized plots and (B) relative check plot yields for individual hybrids.

mineralization, but such data are not available in this case. These data suggest that crop cultivars can also effect the estimate of mineralization in that relative grain yields for first-year check plots varied by at least 20% across hybrids.

Contributions from residual N can be accounted for by measuring the changes in the root-zone NO_3–N before and after the crop, or by basing mineralization estimates on second year check plot N uptake (assuming residual N is removed by the first year check crop). Cabrera and Kissel (1988) provide an example of estimating mineralization via sorghum [*Sorghum bicolor* (L.) Moench] N accumulation, including root N estimates, after adjusting for changes in soil mineral N through 1.2 m and allowing for N inputs from rain and seed. Inputs such as precipitation N (usually <0.1 kg ha^{-1} cm^{-1} depth), irrigation water N (0.1 kg ha^{-1} cm^{-1} depth for water containing 1.0 mg NO^3–N L^{-1}), or small amounts of starter fertilizer N (usually <20 kg N ha^{-1}) can be accounted for by documenting these inputs and subtracting a N value after adjusting for an appropriate N uptake efficiency (e.g., Fox et al., 1989).

Nitrogen leaching losses from check plots are usually relatively small because the crop is growing under N limiting conditions. For the same reason, N losses by denitrification from check plots are frequently ignored since little NO_3 is found in the root zone during the growing season. Volatile N losses from living plant tissue are normally assumed to be insignificant, but any such loss tends to underestimate mineralization. Francis et al. (1993) used ^{15}N fertilizer to quantify volatile N losses from corn and showed that even N deficient plots that received 50 kg N ha^{-1} had volatile N losses of ~30 kg ha^{-1} between silking and harvest. One can only speculate that check plots also should exhibit some loss of N during senescence.

The other precaution with nonfertilized plots is the use of long-term check plots to estimate mineralization for recently fertilized plots. A long-term check plot (not fertilized for 3–5 yr) is likely to yield low estimates of mineralization relative to recently fertilized plots because of the decline in soil organic matter with time in N limited situations. The magnitude of this decline will depend on the level of soil organic matter and the N removed by harvested products vs. N returned in crop residues. Soils with low organic matter and high crop N removals will lose mineralization capacity faster than high organic matter soils experiencing modest crop N removals. Hadas et al. (1989) reported that a 6-yr nonfertilized cotton plot, containing ~800 mg total N kg^{-1} soil, mineralized 43% less N than the corresponding high-fertility plot.

The N uptake by an unfertilized crop remains the best and most practical method to estimate field N mineralization for agronomic crops, provided the precautions noted above are observed.

Soil Nitrate–Nitrogen Monitoring (Uncovered and Covered)

Isolated Intact Soil Tubes

Evaluating field mineralization with sheltered undisturbed fallow soil, in situ mineralization, is a very attractive approach because it avoids the enhanced mineralization associated with soil disturbance during sampling and sample homogenization (drying, crushing, and mixing) used for other methods. The lack of disturbance is especially important for emulation of modern conservation tillage systems (no-till, or chisel-plow), which produce only minimal disruption of the soil, as compared with traditional plow tillage, which mixes the soil each year. Another advantage of intact methods is their potential to incorporate tillage and residue placement side-effects into the field mineralization process. For example, no-tillage culture has been shown to produce a wetter, more dense, and cooler soil with a marked enrichment of nutrients and organic matter in the surface soil. These conditions can all affect field mineralization through the availability and placement of the substrates and through the rate modifying effects of cooler temperatures and wetter moisture conditions. Fallow plots eliminate the need for estimation of root N accumulation in a nonfertilized crop. A potential disadvantage of the sheltered-fallow plot is the disruption of the soil moisture wetting and drying cycles, which normally occur through moisture depletion via evapotranspiration and periodic rewetting with precipitation or irrigation. Soil enclosures also may alter the temperature through greenhouse heating effects associated with limited air exchange.

Covered Fallow Plots. The simplest approach to undisturbed soil mineralization is the covering of an area of soil, to prevent leaching, and periodic sampling of the area after several weeks or months. This approach was described by Powlson (1980) and recently studied by Rice et al. (1987), who covered 1-m^2

areas of no-till or plow-tilled soil across a drainage toposequence at two locations in Kentucky. The covers were 0.1-mm polyethylene sheets on a wooden frame held 20 cm above the soil surface. The covers effectively conserved soil moisture (moisture declined only ~3% during the study) and preserved the somewhat wetter moisture status typical of no-tillage culture (no-till areas contained ~1 to 2% more soil moisture than corresponding plow-till areas). Soil mineral N was measured by soil cores collected beneath the covers after several weeks. The effectiveness of the covers in minimizing leaching and denitrification losses was demonstrated by comparing soil mineral N concentrations in sheltered and unsheltered areas. After 92 d, the no-till sheltered area contained 58 mg of mineral N kg^{-1} of soil, while the unsheltered area contained 26 mg mineral N kg^{-1}; corresponding concentrations in the plow-tilled areas were 84 vs. 85 mg mineral N kg^{-1}. The shelters prevented a leaching loss of >50% of the mineral N in the no-till areas, while N losses were apparently not significant in the plow-tilled sections.

The quantities of N mineralized under sheltered areas for each soil and tillage combination (Table 3–1) show marked differences between soils within a drainage toposequence and between the locations. The average CV for the field mineralization data was ~42%. Part of the location difference may have been due to the fact that a tall fescue (*Festuca arundinacea* Schreber) sod preceded the study at Lexington, which contained ~21 mg N kg^{-1} (C/N of 21, assuming 45% C), while the sod at Princeton contained only 12 mg N kg^{-1} (C/N of ~38). Since this technique measures net mineralization (i.e., mineralization minus immobilization) it is likely that immobilization was enhanced by the low N residues at the Princeton site, which caused a corresponding low mineralization value. Lower mineralization rates were observed with poor drainage, especially with plow tillage, which was probably due to greater denitrification losses in these wetter soils. The data in Table 3–1 also illustrate a marked tillage by drainage interaction that is shown by the enhanced mineralization with plow-tillage on well-drained soils as compared with reduced mineralization on plow-tilled poorly drained soils. This study also evaluated two laboratory mineralization procedures: short-term (2 wk) aerobic mineralization with soil–sand mixtures, and N extracted by 16 h of autoclaving in 0.01 M $CaCl_2$. These standard laboratory methods, using dried and crushed samples, were able to predict differences between soil series, but not tillage differences. A laboratory method using intact cores incubated for ~1 mo, was able to predict relative tillage differences and the interaction between tillage and soil drainage, but was less successful in predicting differences among soil series.

Covered fallow plots were also used by Hadas et al. (1989) on a Vertisol in Israel to estimate mineralization at various depths. These investigators covered a 2- by 1-m area with black 0.03-mm polyethylene and measured NO_3–N through a 1.2-m profile (in 20-cm increments) at five times during an 11-wk period. They also measured temperature and water content and determined the laboratory N mineralization potential (Hadas et al., 1989) with long-term aerobic incubations (as per Stanford & Smith, 1972) in order to predict mineralization and compare predicted vs. measured values. They reported field mineralization of 138 and 78 kg N ha^{-1} from fertilized and nonfertilized plots, respectively. The major difference

Table 3-1. Soil characteristics and N mineralization rates from sheltered-fallow microplots on two Kentucky toposequences under no-tillage culture (Rice et al., 1987).

Toposequence location/Soil name	Drainage class	Soil total N	N mineralization rate	
			No-till	Plow-till
		g N kg⁻¹	mg N kg⁻¹ d⁻¹	
Lexington				
Maury (fine, mixed, mesic Typic Paleudalf)	Well	1.7	0.65	1.48**
Lindside (fine–silty, mixed, mesic Fluvaquentic Eutrochrept)	Moderately well	3.0	1.18	1.35
Lanton (fine–silty, mixed, thermic Cumlic Haplaquoll)	Somewhat poorly	3.4	0.77	0.47*
Princeton				
Zanesville (fine–silty, mixed, mesic Typic Fragiudalf)	Moderately well	0.9	0.30	0.41
Johnsburg (fine–silty, mixed, mesic Aquic Fragiudalf)	Somewhat poorly	0.9	0.17	0.22
Purdy (clayey, mixed, mesic Typic Ochraquult)	Poor	0.8	0.20	0.15

*, **Significance at the 0.05 and 0.01 probability levels, respectively.

between these plots was a greater mineralization in the top 40 cm of the fertilized plot. They also reported that predicted mineralization, laboratory mineralization adjusted for temperature and water content, was only 13 to 26% more than the field value.

Uncovered Fallow Plots. Using uncovered fallow plots to study mineralization is fraught with difficulties, especially in humid climates due to the mobility of NO_3. In semiarid climates without irrigation this method, however, has merit. The approach is straight forward and requires a noncropped plot that is sampled several times during the year to determine the change in mineral N. Since the CV of soil NO_3 data usually averages ~40 to 70%, one should employ an intensive sampling plan to accurately measure the change in soil mineral N.

Intact Soil Cores. Intact soil cores offer the advantages of an undisturbed soil, a clearly defined soil volume, and residue placement appropriate for the tillage system to be studied. The field incubation of intact cores, however, is usually limited to small cores (5–20 cm in diam.) and still requires a method to control excess water inputs in order to avoid NO_3 leaching losses. A novel approach to limit NO_3 leaching from cores was proposed by Schnabel (1983) who placed anion exchange resins below intact cores in order to trap any NO_3 leached through the core. A resin equipped core could then be incubated in the field and exposed to normal field temperature and water conditions. Schnabel also demonstrated that a 2-cm layer of anion exchange resin could reduce the NO_3 concentration of percolating water to 1% of its initial value and that >90% of the adsorbed NO_3 could be recovered after 12 wk of incubation, indicating that N transformations within the resin were negligible.

The use of intact cores with anion exchange resins was studied for 2 yr in Maryland on two contrasting soils representing the Coastal Plain (Galestown loamy sand; sandy, siliceous, mesic Psammentic Hapludult) and Piedmont (Manor silt loam; coarse-loamy, micaceous, mesic Typic Dystrochrept) regions. Four or five seamless-aluminum tubes (5-cm diam.) were driven 25 cm into the surface soil of nonfertilized corn plots, then withdrawn, and the bottom 3 to 4 cm of soil was removed to allow insertion of 30 g (~50 mL) of Ionac A-544, a strong anion exchange resin, which had been placed into a sewn nylon mesh bag. The resin bag was held in place with a small piece of tape, and the open tube and bag were reinserted into the hole for field incubation. At the end of 4 wk the tubes were collected and the soil and resin were analyzed for NO_3-N and NH_4-N by steam distillation after extraction with 2 M KCl. Samples of the surrounding bulk soil (0–25 cm) were also taken before and after the incubation period and analyzed for mineral N. Treatments selected for evaluation at each location included no-tillage vs. plow-till treatments and nonfertilized vs. manure amended plots (Table 3–2). Corn grain yields were measured at the end of the year and are expressed as relative yields (i.e., the yield as a percentage of the non N limited yield at the site as described in Meisinger et al., 1992b).

The soil mineral N data (Table 3–2) show that the manure treatments increased the bulk soil mineral N concentration, but the increases were larger for the soil in the open tubes, particularly for the poultry manure treatments. The dairy manure increased mineral N only a small amount because it contained 4% N (C/N ~12). It should also be noted that the increased NO_3-N plus NH_4-N from

FIELD INDICATORS OF NITROGEN MINERALIZATION

Table 3-2. Soil mineral levels in intact cores, in resin bags below the cores, and in bulk soil as related to tillage and N inputs at two Maryland locations.

Tillage	N added Source	Amount	Bulk soil Start	End	Open tube	Anion resin	Tube + resin	Mineralization rate	Relative grain yield
		kg N ha^{-1}	mg (NO$_3$ + NH$_4$)–N kg^{-1} soil					mg N kg^{-1} d^{-1}	
Coastal Plain, 1987, 51 mm precipitation									
No-till	Fertilizer	none	4	6	12	3	15	0.39	0.41
Plow-till	Fertilizer	none	8	8	11	5	16	0.29	0.57
No-till	Poultry manure	270	27	21	25	12	37	0.36	0.83
Plow-till	Poultry manure	270	35	35	50	28	78	1.54	1.00
Coastal Plain, 1988, 9 mm precipitation									
No-till	Fertilizer	none	12	12	14	1	15	0.11	0.56
Plow-till	Fertilizer	none	14	14	19	2	21	0.26	0.69
No-till	Poultry manure	270	28	26	35	3	38	0.37	0.90
Plow-till	Poultry manure	270	31	33	54	6	60	1.07	0.92
Piedmont, 1987, 55 mm precipitation									
No-till	Fertilizer	none	10	12	13	2	15	0.17	0.36
Plow-till	Fertilizer	none	12	11	15	3	18	0.20	0.32
No-till	Dairy manure	80	12	14	17	5	22	0.33	1.00
Plow-till	Dairy manure	80	13	15	20	6	26	0.43	0.82
Piedmont, 1988, 12 mm precipitation									
No-till	Fertilizer	none	11	14	23	1	24	0.59	0.70
Plow-till	Fertilizer	none	12	14	19	1	20	0.36	0.63
No-till	Dairy manure	80	14	21	34	2	36	1.00	0.90
Plow-till	Dairy manure	80	13	23	25	2	27	0.64	0.90

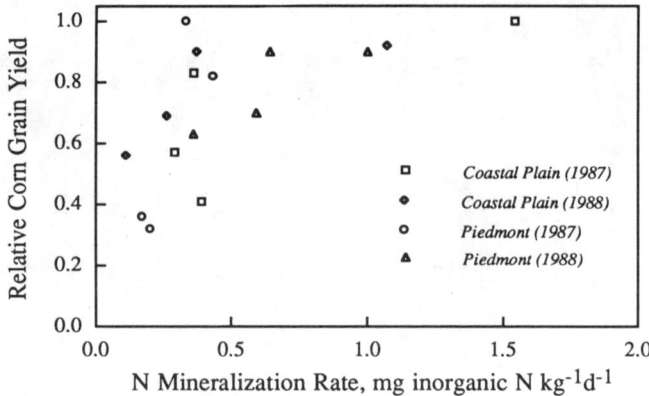

Fig. 3–2. Nitrogen mineralization rate for two locations in Maryland and relative corn grain yields.

the poultry manure was smaller in the no-till treatment, where it was surface applied, than in the plow-till treatment. This is probably due to significant NH_3 volatilization losses in no-till compared with insignificant losses for the incorporated manure in plow-tillage. The anion exchange resin accumulated significant quantities of NO_3–N in the loamy sand soil of the Coastal Plain for the 55-mm precipitation in 1987, presumably this NO_3 would have been lost out of the sample zone if unprotected cores had been used. Smaller quantities of NO_3 were trapped by the resin in the silt loam soil in the Piedmont, and in 1988 at both sites when rainfall was minimal during the field incubation. The CVs among cores within a plot for the open tubes averaged 58% and the CVs among resin cores within a plot averaged 48%. Thus, the variability among tubes for the resin-core method was about the same as encountered for soil NO_3–N. The relationship between field mineralization rate (mineral N in tube plus resin divided by the incubation time) and relative corn grain yield (Fig. 3–2) shows that nearly maximum yields were obtained with mineralization rates of ~0.8 mg NO_3–N plus NH_4–N kg^{-1} d^{-1} or more. It is important to note that these mineralization rates are not based on an annual or seasonal time interval, but are for the month between planting and the V6 growth stage (Ritchie et al., 1986) and contain the mineralization burst from the recent manure additions.

The resin-core technique has not been adequately evaluated in the field. The data above indicate that it has considerable potential in humid climates, but more work is needed to determine the limits of its usefulness in a wider range of soils and climates.

Mineralization Bags. This method was originally developed by Eno (1960) and consists of enclosing a soil sample in a polyethylene bag and replacing it in the field for a period of time, usually a month or more. A common modification to the basic method uses undisturbed blocks or cylinders of soil (Poovarodom et al., 1988; Pastor et al., 1984). Advantages are that this approach eliminates NO_3 leaching and permits incubation under field temperature conditions and facilitates precise recovery of the original sample, which reduces field variability compared with estimating mineralization as the difference between two independent soil samples. Disadvantages include the inability to mimic soil moisture conditions, particularly wetting and drying cycles. The possibility of disturbed aeration also exists due to O_2 and CO_2 gas diffusion limitations through the polyethylene (Bremner & Douglas, 1971), although Westermann and Crothers (1980) reported no restrictions in CO_2 diffusion rates. They believed any restrictions in CO_2 diffusion had little or no effect on N transformations.

This method is frequently used in forestry and ecological research where large perennial species are being studied under low N input situations. Poovarodom et al. (1988) used polyethylene bags to enclose disturbed (sieved) and undisturbed blocks of soil (~750 cm³) from two soil types in the New Jersey Pine Barrens Region. Monthly samples were collected along a 20-m transect at 2-m intervals to a depth of 15 cm for each soil type for 1 yr to estimate seasonal mineralization values. The average monthly mineralization rates for these soils varied seasonally with maximum rates in the summer, 4 to 6 kg N ha^{-1} mo^{-1}; and minimum rates in the winter, <1 kg N ha^{-1} mo^{-1}. Disturbed vs. undisturbed soil had no statistically significant effect on annual N mineralization, but disturbed soil usually gave higher mineralization values throughout the spring. The average CVs among samples within the 20-m transect was 68%. The final yearly mineralization estimates were 38 kg N ha^{-1} (4.5% of the total N) and 53 kg N ha^{-1} (2.5% of the total N) for these very acid soils (pH ~4.0).

Buried bags were also evaluated by Westermann and Crothers (1980) as a method to characterize N mineralization under field conditions for corn and potato (*Solanum tuberosum* L.). They removed 0- to 45-cm or 0- to 30-cm depth cores (5.7 cm in diam.) from fields in Idaho, then sieved, combined, and mixed the soil and added water to ~75% of field capacity. The moist homogenized soil was then placed in 6-mm thick cylindrical polyethylene bags (5-cm diam. by 30- or 45-cm long) and replaced in the original core holes for field incubation. They found that the soil NO_3–N concentration in the polyethylene bags was similar to that in irrigated fallow plots, provided adjustments were made for soil moisture. The moisture correction involved equating the relative mineralization rate or relative soil water content, expressed as a percentage of field capacity (e.g., at 75% of field capacity mineralization would be 75% of optimum; Stanford & Epstein, 1974). Furthermore, N uptake by the corn and potato crops were similar to those predicted from the buried bags. They also reported a Q_{10} (the relative change in microbial activity associated with a 10°C increase in temperature) for mineralization of ~2, between 10 and 30°C, as previously suggested by Stanford et al. (1973). The CVs among incubation bags within a field was only 8%, due to the homogenizing and mixing of the sample before field incubation. The investigators concluded that the buried bag technique provided a valid alternative method to estimate mineralization under field conditions.

APPROACHES FOR RELATIVE ESTIMATES

Presidedress Nitrate Test Estimates for Nitrogen Sufficiency

The presidedress nitrate test estimate (PSNT) is based on a timely monitoring of the in situ field N mineralization process and is primarily a relative index of the N mineralization intensity, i.e., a NO_3–N concentration is used rather than the mass of NO_3–N. The PSNT measures the NO_3–N concentration in the surface 30 cm of soil when corn is 15- to 30-cm tall. The PSNT was first proposed by Magdoff et al. (1984).

The PSNT takes full advantage of the yearly rhythms in the soil NO_3–N pool that result from seasonal differences in the NO_3 production processes vs. NO_3 loss processes. In humid temperate climates a typical silt loam soil growing corn, will generally have surface soil NO_3–N contents that are lowest in late winter and early spring due to recent winter leaching and denitrification associated with high soil moisture levels, and low mineralization due to cold temperatures. Common NO_3–N concentrations in soil after winter leaching (or excess irrigation) would be 5 to 10 mg NO_3–N kg^{-1} (Broadbent & Carlton, 1979; Hahne et al., 1977; Olsen et al., 1970; Sarrantonio & Scott, 1988). In spring and early summer the NO_3–N content increases due to greater mineralization resulting from warming temperatures and spring tillage, availability of fresh organic matter from crop residues or manure, little or no corn N uptake, and lower leaching and denitrification losses due to lower soil moisture levels resulting from increased evapotranspiration (ET).

Nitrate-N concentrations in nonfertilized soil may increase two- to six-fold in late spring depending on recent additions of crop residues or manure (Fox et al., 1989; Magdoff et al., 1984; Magdoff, 1991a,b; Meisinger et al., 1992b). This late spring increase in NO_3–N from in situ mineralization is the underlying principle for the PSNT. In summer, soil NO_3–N levels decline dramatically as soil N mineralization is unable to keep pace with corn N uptake. In the fall, there is usually a small increase in the soil NO_3 pool due to: corn senescence, continued mineralization under moderate temperatures, and low leaching and denitrification losses due to low soil moisture resulting from high summer ET. Figure 3–3

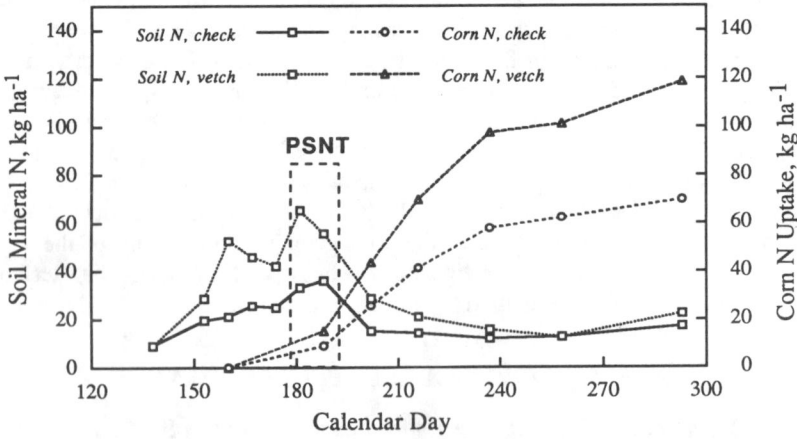

Fig. 3–3. Soil mineral N content and corn N uptake throughout the growing season in a Central New York silt loam soil previously receiving no fertilizer or after a hairy vetch winter cover crop (Sarrantonio & Scott, 1988).

summarizes data from Sarrantonio and Scott (1988) that exemplifies these patterns for central New York conditions, along with the estimated corn N uptake.

The PSNT is not a universal N mineralization test. It has been accepted by most New England and Mid-Atlantic states, and by some Midwest states (Iowa, Wisconsin, and Michigan), but is still being evaluated in most other Midwestern states and in the upper South. The PSNT has several limitations, namely: used only for corn, preferably used with sidedress fertilizer management, requires a strict time of sampling during a short period, requires rapid analysis, is affected by excess rainfall and cold temperatures, and has limited usefulness for N deficient sites. Nevertheless, once these limitations are recognized and the test carefully employed it has proven to be very useful as a N sufficiency index. Soils with NO_3–N concentrations >20 to 25 mg N kg^{-1} have been associated with little or no response to additional fertilizer N (Blackmer et al., 1989; Fox et al., 1989; Meisinger et al., 1992a). States using the PSNT have commonly found that it has saved farmers an average of 33 to 56 kg ha^{-1} of fertilizer N (Woodward et al., 1993). These circumstances most commonly occur on fields previously treated with manure or sludge.

Chlorophyll Meter

The crop N uptake approach provides an estimate of mineralization during the growing season, but as noted earlier, the short-term dynamics of the N cycle may be equally or more important to producers. Tissue analysis for N content or chlorophyll meters can be used to provide a qualitative indication of mineralization during the growing season (Peterson et al., 1993; Schepers et al., 1992a). This is accomplished by comparing tissue N concentrations or meter readings from an area of the field receiving adequate N fertilizer with those where N availability is in question. Critical levels that characterize crop N status are questionable for tissue N concentration and not available for chlorophyll meter readings, which is why calculation of a sufficiency index is probably the best approach to assessing mineralization while the crop is growing.

Another approach to estimate the amount of N that can realistically be attributed to mineralization of soil organic matter and organic waste products is to monitor relative crop N status or N uptake over time (Schepers et al., 1992b). This approach typically requires an extended commitment of time and is probably best suited to research situations where ongoing technical expertise is available. Experimental procedures will differ depending on the situation and how the results will be applied to producer fields. Background information provided by soil test data and chemical analyses of waste products are essential in all cases. This type of mineralization research typically provides vital information about when and how rapidly mineralization occurs by qualitatively comparing crop N uptake or crop N status between various treatments. An example of this approach has been demonstrated at the Nebraska Management Systems Evaluation Area (MSEA) project where four types of agricultural wastes are applied to land planted to corn. Waste application rates were adjusted so that P contained in the waste was equivalent to the amount of P removed in the grain during a 2-yr period. As such, the wastes were applied in alternate years. Because differing amounts of manure were applied to provide the same amount of total P, total N application rates for the beef, sheep,

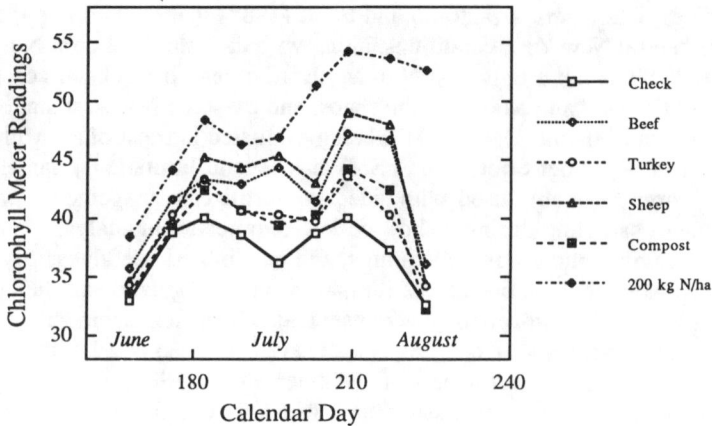

Fig. 3–4. Chlorophyll meter readings for irrigated corn in 1993 at Shelton, NE, for various types of animal waste.

turkey, and composted paunch manures were 228, 174, 113, and 142 kg N ha^{-1}, respectively. Chlorophyll meter data showed that some types of spring applied and incorporated manures showed little net N mineralization compared with nonfertilized plots, but later demonstrated various amounts of net mineralization (Fig. 3–4). The differences in crop N status at times during the growing season indicated apparent differences in N mineralization rates or N applied for the various manures.

Electro-Ultra Filtration Methods

Some cropping systems can be adversely impacted by excess N mineralization at critical times during the growing season. The problem with sugar beet (*Beta vulgaris* L.) production is a widely recognized example since excess availability of soil NO_3 late in the growing season typically results in high levels of dry matter production with low sugar content. As such, producers with low sugar content beets are penalized because they must pay for processing larger volumes of material that may yield lower amounts of sugar per unit area of land.

The above scenario was recognized as a serious problem in Europe in the early 1970s. Efforts to use soil testing for residual soil N to identify potential problem sites prior to planting were only marginally successful. Failures to detect problem situations were usually attributed to sites with a history of manure application or where climatic and soil conditions promoted above normal mineralization. The problem with excess soil NO_3 for sugar beet production comes to bear in August and early September when N uptake should ideally become limiting.

The realization that producers needed ways to identify high mineralizing soils prompted some facets of the European sugar beet industry to develop special laboratories that focused on procedures that quantified potentially mineralizable N (PMN) in soils intended for beet production. One of these laboratories uses the electro-ultra filtration (EUF) extraction procedure to prepare a liquid sample for NO_3 and NH_4 analyses, plus a number of other elements. The EUF procedure places a voltage across electrodes in a soil solution, which fractures some of the soil chemical bonds and causes a migration of the anion and cation species to the electrodes (Saint-Fort et al., 1993). Ionic species that congregate near the electrodes are vacuum extracted through millipore filters located adjacent to each electrode. The two leachate fractions are then recombined for chemical analyses. Solution temperature and voltage are controlled to extract a series of two samples (EUF_1 and EUF_2) that are considered *readily available* and *reserve* N components. Data interpretation and procedures used to make fertilizer recommendations are probably the most advanced for sugar beet production, but still apparently require adjustments for climatic regions or soil types.

Because the EUF extraction procedure requires 35 min per sample and the equipment is very expensive, a number of other techniques have been proposed as possible substitutes. Saint-Fort et al. (1990) compared six chemical indices of N mineralization using a select group of U.S. Soil Conservation Service benchmark soils and showed the best correlation ($r^2 = 0.91$) between EUF extractable-N and autoclave labile-N. Multiple regression analysis showed that the combination of soil organic matter content and growing degree days was better correlated with N uptake by corn grown on long-term check plots at five locations in Nebraska than was either EUF extractable-N and autoclave labile-N.

The most unique feature of the EUF approach for assessing PMN relates to when the samples are collected. Rather than collecting samples prior to planting as is common for many row crops, samples for EUF analysis are typically collected from the surface soil in August of the year prior to when sugar beets will be grown. This strategy relies on N uptake by the previous crop to have depleted soil residual N to a minimal level by late summer. Therefore, occurrence of higher than expected levels of EUF-N signals potential problems for subsequent sugar beet production.

The effect of excess soil N availability can also impact small grain production by inducing lodging, which makes harvesting more difficult and can reduce yields. Even when lodging is not a problem, the quality of certain small grains can be impaired by excess N. Such is the case with malting barley (*Hordeum vulgare* L.) where high protein grain interferes with fermentation. Mineralization contributes to these situations by contributing to the pool of plant available N in soil. This introduces the possibility for proper management of fertilizer N rates by crediting for mineralization if it can be quantified. Even then, field variability makes it difficult to properly credit mineralization unless variable rate application technology is available to compensate for field variability in soil N status caused by other N sources.

OVERVIEW

Field measurements of N mineralization provide a degree of reality that may be difficult to achieve with either laboratory procedures or computer simulations. The various field approaches to estimate mineralization have strengths and weaknesses that should be considered before conducting a study. The primary advantage of field mineralization techniques have over laboratory procedures is that they integrate many soil, climatic, and management factors over time. Field techniques that use crop N uptake as a measure of mineralization have the additional advantage that the extensive root system of plants integrates a larger area than can be achieved with discrete soil samples. The shortcoming of field mineralization techniques is that the dynamics of soil temperature, water content, chemical properties, microbial activity, and plant response are seldom adequately documented so that the data can be used to simulate mineralization under other conditions. Field estimates of N mineralization tend to be site specific for a cropping system and climatic situation.

REFERENCES

Blackmer, A.M., D. Pottker, M.E. Cerrato, and J. Webb. 1989. Correlations between soil nitrate concentrations in late spring and corn yields in Iowa. J. Prod. Agric. 2:103–109.

Bremner, J.M. 1965. Nitrogen availability indexes. p. 1324–1345. In C.A. Black et al. (ed.) Methods of soil analysis. Part 2. Agron. Monogr. 9. ASA, CSSA, and SSSA, Madison, WI.

Bremner, J.M., and L.A. Douglas. 1971. Use of plastic films for aeration in soil incubation experiments. Soil Biol. Biochem. 3:289–296.

Broadbent, F.E., and A.B. Carlton. 1979. Field trials with isotopically labeled nitrogen fertilizer. p. 1–41. In D.R. Nielsen and J.G. MacDonald (ed.) Nitrogen in the environment. Vol. I. Nitrogen behavior in field soil. Academic Press, New York.

Cabrera, M.L., and D.E. Kissel. 1988. Evaluation of a method to predict nitrogen mineralization from soil organic matter under field conditions. Soil Sci. Soc. Am. J. 52:1027–1031.

Eno, C.F. 1960. Nitrate production in the field by incubating the soil in polyethylene bags. Soil Sci. Soc. Am. Proc. 24:277–279.

Fox, R.H., G.W. Roth, K.V. Iverson, and W.P. Piekielek. 1989. Soil and tissue nitrate tests compared for predicting soil nitrogen availability to corn. Agron. J. 81:971–974.

Francis, D.D., Schepers, J.S., and M.F. Vigil. 1993. Post-anthesis nitrogen loss from corn plants. Agron. J. 85:659–663.

Griffin, G.F., and A.F. Laine. 1983. Nitrogen mineralization in soils previously amended with organic wastes. Agron. J. 75:124–129.

Hadas, A., A. Feigin, S. Feigenbaum, and R. Portoy. 1989. Nitrogen mineralization in the field at various soil depths. J. Soil Sci. 40:131–137.

Hahne, H.C., W. Kroontje, and J.A. Lutz. 1977. Nitrogen fertilization: I. Nitrate accumulation and losses under continuous corn cropping. Soil Sci. Soc. Am. J. 41:562–567.

Magdoff, F.R. 1991a. Understanding the Magdoff pre-sidedress nitrate test for corn. J. Prod. Agric. 4:297–305.

Magdoff, F.R. 1991b. Field nitrogen dynamics: Implications for assessing N availability. Commun. Soil Sci. Plant Anal. 22:1507–1517.

Magdoff, F.R., D. Ross, and J. Amdon. 1984. A soil test for nitrogen availability to corn. Soil Sci. Soc. Am. J. 48:1301–1304.

Meisinger, J.J., V.A. Bandel, J.S. Angle, B.E. O'Keefe, and C.M. Reynolds. 1992a. Presidedress soil nitrate test evaluation in Maryland. Soil Sci. Soc. Am. J. 56:1527–1532.

Meisinger, J.J., F.R. Magdoff, and J.S. Schepers. 1992b. Predicting N fertilizer needs for corn in humid regions: underlying principles. p. 7–27. In B.R. Bock and K.R. Kelley (ed.) Predicting N fertilizer needs for corn in humid regions. Bull. Y-226. Natl. Fertil. and Environ. Res. Ctr., Muscle Shoals, AL.

Olsen, R.J., R.F. Hensler, O.J. Attoe, S.A. Witzed, and L.A. Peterson. 1970. Fertilizer nitrogen and crop rotation in relation to movement of nitrate nitrogen through soil profiles. Soil Sci. Soc. Am. Proc. 34:448–452.

Paster, J., J.D. Aber, and C.A. McClaugherty. 1984. Above-ground production and N and P cycling along a nitrogen mineralization gradient on Blackhawk Island, Wisconsin. Ecology 65:256–268.

Peterson, T.A., T.M. Blackmer, D.D. Francis, and J.S. Schepers. 1993. Using a chlorophyll meter to improve N management. Coop. Ext. Serv. NebGuide G93-1171A. Univ. Nebraska, Lincoln.

Poovarodom, S., R.L. Tate, and R.A. Bloom. 1988. Nitrogen mineralization rates of the acidic, xeric soils of the New Jersey pinelands: Field rates. Soil Sci. 145:257–263.

Powlson, D.S. 1980. Effect of cultivation on the mineralization of nitrogen in soil. Plant Soil 57:151–153.

Rice, C.W., J.H. Grove, and M.S. Smith. 1987. Estimating soil net nitrogen mineralization as affected by tillage and soil drainage due to topographic position. Can. J. Soil Sci. 67:513–520.

Ritchie, S.W., J.J. Hanway, and G.O. Benson. 1986. How a corn plant developes. Iowa State Univ. Coop. Ext. Serv. Spec. Rep. 48. Ames.

Saint-Fort, R, K.D. Frank, and J.S. Schepers. 1990. Role of nitrogen mineralization in fertilizer recommendations. Commun. Soil Sci. Plant Anal. 21(13–16):1945–1958.

Saint-Fort, R., J.S. Schepers, and C.S. Holzhey. 1993. Comparison of potentially mineralizable nitrogen using electro-ultrafiltration and some other procedures. J. Environ. Sci. Health. A28(6):1171–1183.

Sarrantonio, M., and T.W. Scott. 1988. Tillage effects on availability of nitrogen to corn following a winter green manure crop. Soil Sci. Soc. Am. J. 52:1661–1668.

Schepers, J.S., T.M. Blackmer, and D.D. Francis. 1992a. Predicting N fertilizer needs for corn in humid regions: using chlorophyll meters. p. 103–114. In B.R. Bock and K.R. Kelley (ed.) Predicting N fertilizer needs for corn in humid regions. Bull. Y-226. Natl. Fertil. and Environ. Res. Ctr. Muscle Shoals, AL.

Schepers, J.S., D.D. Francis, M. Vigil, and F.E. Below. 1992b. Comparison of corn leaf nitrogen and chlorophyll meter readings. Commun. Soil Sci. Plant Anal. 23(7–20):2173–2187.

Schnabel, R.R. 1983. Measuring nitrogen leaching with ion exchange resin: A laboratory assessment. Soil Sci. Soc. Am. J. 47:1041–1042.

Smith, S.J., L.B. Young, and G.E. Miller. 1977. Evaluation of soil nitrogen mineralization potential under modified field conditions. Soil Sci. Soc. Am. J. 41:74–76.

Soil Science Society of America. 1987. Glossary of soil science terms. SSSA, Madison, WI.

Stanford, G., J.N Carter, D.T. Westerman, and J.J. Meisinger. 1977. Residual nitrate and mineralizable soil nitrogen in relation to nitrogen uptake by irrigated sugar beets. Agron. J. 69:303–308.

Stanford, G., and E. Epstein. 1974. Nitrogen mineralization-water relations in soils. Soil Sci. Soc. Am. Proc. 38:103–107.

Stanford, G., M.H. Frere, and D.H. Schwaninger. 1973. Temperature coefficient of soil nitrogen mineralization. Soil Sci. 115:321–323.

Stanford, G., and S.J. Smith. 1972. Nitrogen mineralization potentials of soils. Soil Sci. Soc Am. Proc. 36:465–472.

Westermann, D.T., and S.E. Crothers. 1980. Measuring soil nitrogen mineralization under field conditions. Agron. J. 72:1009–1012.

Woodward, M.D., V.A. Bandel, and B.R. Bock. 1993. Summary of soil nitrate tests for corn, 1993 State Surveys. Tennessee Valley Authority Misc. Publ. Tennessee Valley Authority, Muscle Shoals, AL.

4 Linking Nitrogen Mineralization and Plant Nitrogen Demand with Thermal Units

C. Wayne Honeycutt

USDA-ARS
New England Plant, Soil, and Water Laboratory
University of Maine
Orono, Maine

Plant recovery of fertilizer N is reportedly 50% or less for several agronomic crops (Keeney, 1982; Hallberg, 1987). This unrecovered N constitutes a potential contributor to groundwater contamination, eutrophication, acid rain, global warming, and farm insolvency. A primary goal of predicting N mineralization from soils and other organic N sources is to increase the overall efficiency of N use in crop production.

Stanford and Smith (1972) introduced the concept of N mineralization potential as a quantitative estimate of soil N mineralization. This approach has been determined successful in several field studies (Stanford et al., 1977; Marion et al., 1981; Griffin & Laine, 1983). Others have reported the method to overpredict field N mineralization by 67 to 343% (Cabrera & Kissel, 1988).

Important components of the Stanford and Smith (1972) approach include modification of a first-order rate constant for soil temperature and water conditions (Stanford et al., 1973; Stanford & Epstein, 1974). Mineralization rate constants are generally assumed to change two-fold for each 10°C change in temperature (i.e., Q_{10} = 2.0) (Stanford et al., 1973); however, wide variations in the Q_{10} value are reported. Ross and Bridger (1978) reported Q_{10} values for N mineralization to range from 1.7 to 3.8. Tabatabai and Al-Khafaji (1980) reported Q_{10} to range from 2.5 to 4.1. Values calculated from the Stanford et al. (1973) data indicate Q_{10} varied between 1.2 and 3.0.

The significance of this variation in Q_{10} becomes evident upon considering the range in soil temperatures normally encountered under field conditions during a growing season. For soil temperatures ranging from 5 to 35°C, a Q_{10} of 2.0 translates into an eight-fold difference in the mineralization rate constant, while a Q_{10} of 3.0 translates into a 27-fold difference in this constant. Consequently, choice of temperature coefficient can have a dramatic impact on N mineralization prediction accuracy with this procedure.

An alternative to the kinetic approach for predicting N mineralization employs the concept of thermal units (degree days) (Honeycutt et al., 1988).

Copyright © 1994 Soil Science Society of America, 677 S. Segoe Rd., Madison, WI 53711, USA.
Soil Testing: Prospects for Improving Nutrient Recommendations, SSSA Special Publication 40.

Particular advantages of thermal units include (i) no preconceived notion of the nature of the N mineralization process (e.g., zero-order vs. first-order kinetics; single vs. multiple labile organic N pools), (ii) no assumed Q_{10} value, (iii) a common unit to compare N mineralization with plant N demand, and (iv) grower familiarity with the thermal unit concept. Notable disadvantages of the thermal unit approach are (i) it is an empirical relationship, and (ii) it currently requires conducting an incubation study.

A primary objective of this chapter is to summarize the research conducted on predicting N mineralization from soils and soil amendments with thermal units. A second objective is to detail a proposed procedure for linking N mineralization and plant N demand with thermal units, in hope that N use efficiency in crop production will be enhanced through this relationship. Because thermal units are founded on describing plant growth, this foundation is first briefly reviewed in the following section.

THERMAL UNITS AND PLANT GROWTH

Reamur (1735) was reportedly one of the first to document that plant development is more closely related to accumulated temperature than to time alone (Neild & Seeley, 1977). Several studies have since demonstrated the utility of thermal units for describing plant growth and development (Gilmore & Rogers, 1958; Cross & Zuber, 1972; Klepper et al., 1982; Russelle et al., 1984).

Much of the early work on plant growth and thermal unit relations was based on Lehenbauer's (1914) data. These observations demonstrated distinct minima and optimal temperatures for corn (*Zea mays* L.) seedling growth rates (Fig. 4–1). Relating temperature to plant growth incorporates these observations into the

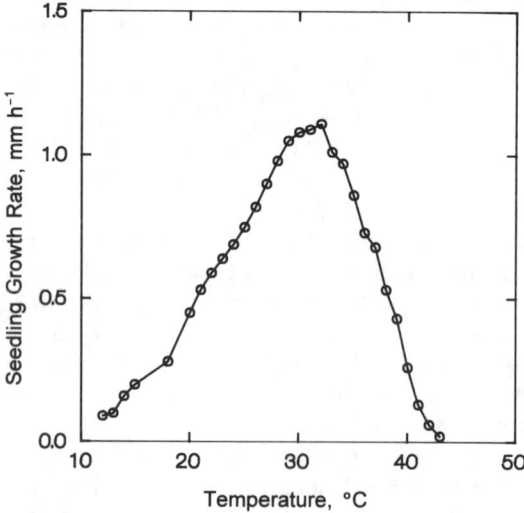

Fig. 4–1. Corn seedling growth rate in relation to temperature (Lehenbauer, 1914).

thermal unit calculation. Temperatures below the minimum threshold for plant growth are considered equal to the minimum (i.e., zero thermal units are accumulated), and superoptimal temperatures are penalized by the amount they exceed the optimum (Gilmore & Rogers, 1958).

THERMAL UNITS AND DECOMPOSITION

A general relationship between soil microbial activity and temperature is strikingly similar to that previously shown for plant growth rate and temperature (Fig. 4–1; Atlas, 1988; Paul & Clark, 1989). The utility of thermal units for predicting decomposition and N mineralization in soils was examined based on this demonstrated utility for describing plant growth (Honeycutt et al., 1988).

One of the early studies to relate soil microbial processes to thermal units was conducted by Miller (1974). Carbon dioxide evolution from sewage sludge-amended soils was significantly related to monthly degree days in that study. Gilmour et al. (1977) presented theoretical data relating thermal units to the half-life of organic C. Andrén and Paustian (1987) used thermal units to describe barley (*Hordeum distichon* L. cv. Gunilla) straw mass loss in field litter bags.

Papermill sludge decomposition also was described with thermal units (Honeycutt et al., 1988). A wide range in cumulative C mineralization was observed with temperature for a given day following sludge application (Fig. 4–2). This range was greatly reduced upon employing thermal units (Fig. 4–3). As many as four decomposition phases, each characterized by different substrate C mineralization rates, were identified during the course of sludge decomposition. Consequently, thermal units described sludge decomposition

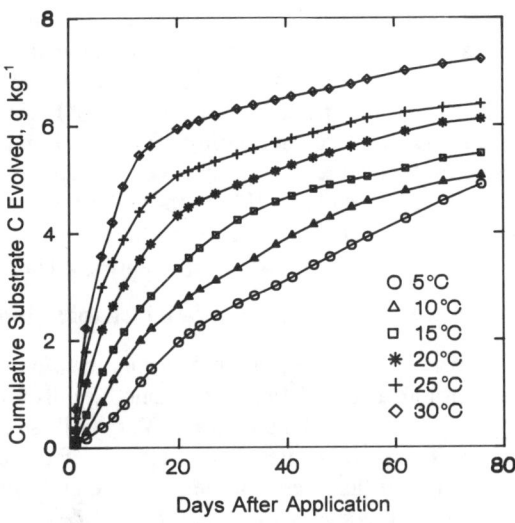

Fig. 4–2. Cumulative substrate C evolved related to time for six incubation temperatures (Honeycutt et al., 1988).

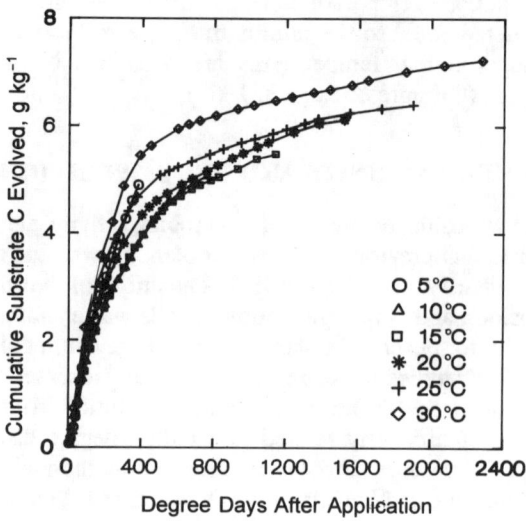

Fig. 4–3. Cumulative substrate C evolved related to thermal units for six incubation temperatures (Honeycutt et al., 1988).

despite variable mineralization rates resulting from different temperatures and apparently changing substrate C labilities (Honeycutt et al., 1988).

THERMAL UNITS AND NITROGEN MINERALIZATION

Circumstantial evidence exists that thermal units can account for temperature effects on N mineralization under fluctuating temperatures. Campbell et al. (1971) observed similar rates of net N mineralization in a soil subjected to diurnally fluctuating temperatures of 14 to 3°C as observed at a constant temperature of 8.5°C. This observation lends support to the thermal unit concept because the total thermal load was equal for both treatments. Using a 0°C baseline for thermal unit accumulation, soil in both treatments accumulated 136 degree days during the 16-d incubation period. Stanford et al. (1975) reported no effect of four different temperature sequences on N mineralization from three agricultural soils. Again, the total thermal loads were equal for each temperature sequence examined.

Nitrogen Mineralization from Soil Organic Matter

The utility of thermal units for predicting N mineralization from soil organic matter was assessed through a compilation of published results for NO_3^- formation (Honeycutt et al., 1991). These results therefore reflect both ammonification and nitrification processes. Three procedures were used to determine this utility: (i) comparing the significance of the interaction terms when NO_3^- concentration was modeled vs. days and when modeled vs. degree days; (ii) determining which variable, day or degree day, was selected first in a stepwise regression analysis; and (iii) graphical comparisons of the range in NO_3^- concentration when plotted against days and degree days. Data from three countries, seven states, and 21 soil series were assessed by each approach (Honeycutt et al., 1991).

In the first approach, if temperature × day and temperature × day^2 are significant in a quadratic model, but temperature × degree day and temperature × (degree day)^2 are not significant, this would indicate the curves differ in shape when expressed vs. days, but not when expressed vs. degree days. In the second approach, the variable selected first in a stepwise regression analysis when both day and degree day were modeled against NO_3^- concentration was interpreted to indicate which of the two independent variables was the strongest predictor. The third approach was conducted to visually compare the NO_3^- vs. day and NO_3^- vs. degree day relationships.

Approximately 26% of those cases exhibiting significant temperature × day and temperature × day^2 interactions also exhibited nonsignificant interactions of temperature × degree day and temperature × (degree day)^2 (Table 4–1). These cases indicate different relationships between NO_3^- concentration and days, but similar relationships between NO_3^- concentration and degree days across temperature treatments. This approach, however, could not be tested for ~67% of the cases due to either nonsignificant interactions of temperature with day and day^2 or the presence of a singular matrix in the analysis (Table 4–1). Use of mean values reported in the literature restricted analysis with the first approach because limited observations can produce the singular matrix. Only 7% of the data files observed to possess significant temperature × day or temperature × day^2 interactions also exhibited significant temperature × degree day or temperature × (degree day)^2 interactions. The second approach for assessing thermal unit utility indicated degree days to be a stronger predictor of NO_3^- concentration than days in 24 out of 27 (89%) data files examined (Table 4–1).

Although more subjective and difficult to quantify, graphical comparisons indicated substantial reductions in NO_3^- concentration variability when assessed with thermal units for ~88% of the data files examined. An example of the reduction in variability afforded by thermal units is provided in Fig. 4–4 and 4–5. Nitrate concentrations at Day 35 ranged from 56 to 144 mg kg^{-1} (Fig. 4–4). Plotting the same data against degree days greatly reduced the range in NO_3^- concentration, with all treatments represented by essentially the same curve (Fig. 4–5).

An important question is: How universal are these relationships? This question is addressed in a following section on factors affecting prediction accuracy.

Nitrogen Mineralization from Organic Amendments

Papermill Sludge

Thermal units were first tested for predicting N mineralization with a papermill sludge (Honeycutt et al., 1988). Sludge addition induced a period of net N immobilization at each of six temperature treatments (Fig. 4–6). Nitrogen immobilization was still greater than mineralization at the close of the incubation study for the 5°C treatment. From 12 to 60 d were required for commencement of net N mineralization in the 30 to 10°C treatments (Fig. 4–6). Examination of the same data in relation to thermal units, however, shows the 10 to 25°C treatments to fall essentially on the same curve (Fig. 4–7). Net N mineralization began between 600 and 684 degree days for these four temperatures (Table 4–2).

Table 4-1. Summary of statistical analyses of studies used to assess NO_3^- formation in relation to thermal units (adapted from Honeycutt et al., 1991).

Reference	Data file†	Significance of temperature interaction with		Stepwise regression variable selected first	
		Day	Degree day	Day	Degree day
Doel et al. (1990)	1	*	—‡		X
	2	NS	—		X
	3	*	—		X
Frederick (1956)	1	NS	NS		X
	2	*	NS		X
	3	—	—		X
Justice & Smith (1962)	1	NS	NS		X
	2	NS	NS		X
	3	NS	NS		X
Kowalenko & Cameron (1976)	1	**	NS		X
	2	**	NS		X
	3	**	**		X
Myers (1975)	1	**	**		X
Panganiban (1925)	1	—	—		X
	2	—	—		X
	3	—	—		X
Parker & Larson (1962)	1	NS	NS		X
	2	NS	NS		X
	3	NS	NS		X
	4	*	**		X
	5	**	NS		X
Russel et al. (1925)	1	*	NS		X
	2	NS	NS		X
Tyler et al. (1959)	1	**	NS		X
	2	NS	NS	X	
	3	*	NS	X	
	4	NS	NS	X	

*,** Significant at $\alpha = 0.10$ and 0.05, respectively; NS = not significant at $\alpha = 0.10$.
† Individual data files were constructed for each soil series, soil amendment, soil pH, and soil water treatment.
‡ —, could not be determined due to the presence of a singular matrix in the analysis.

Soil temperatures in the climatic region studied do not commonly reach 30°C. Consequently, the marked deviation in these results for the 30°C treatment may indicate artifactitious promotion of microbial populations with differing efficiency or N requirement than those normally prevalent in the soil studied. This observation also underscores the possibility that the 35°C temperature recommended for N mineralization potential determination (Stanford & Smith, 1972) may not be appropriate for all soils.

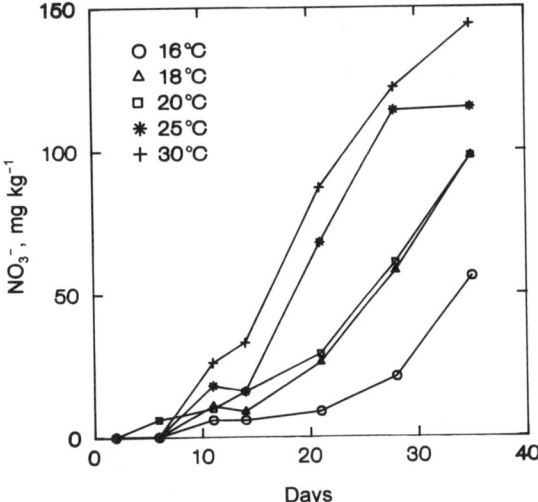

Fig. 4–4. Nitrate concentration at five temperatures in an unamended soil related to time (Parker & Larson, 1962).

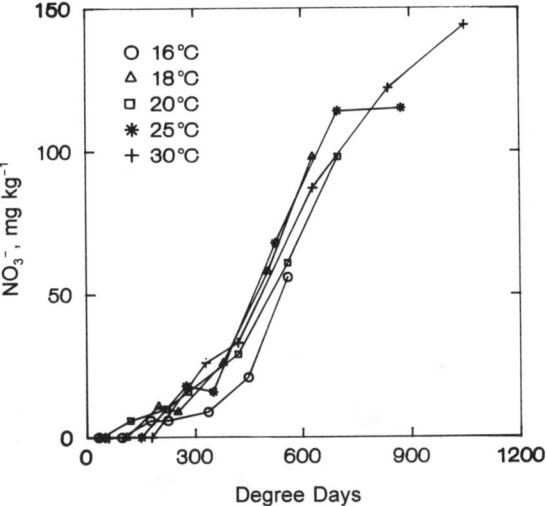

Fig. 4–5. Nitrate concentration at five temperatures in an unamended soil related to thermal units (Parker & Larson, 1962).

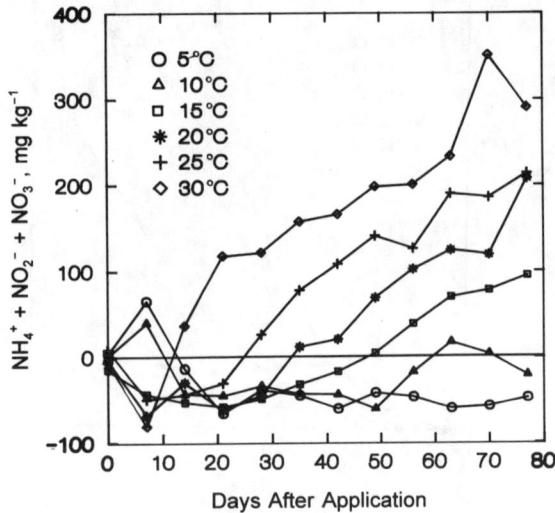

Fig. 4–6. Concentration of $NH_4^+ + NO_2^- + NO_3^-$ in a papermill sludge-amended soil minus that in a control soil related to time for six incubation temperatures (Honeycutt et al., 1988).

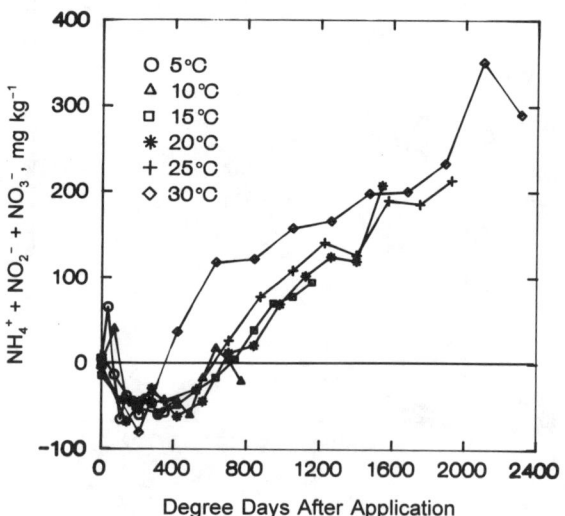

Fig. 4–7. Concentration of $NH_4^+ + NO_2^- + NO_3^-$ in a papermill sludge-amended soil minus that in a control soil related to thermal units for six incubation temperatures (Honeycutt et al., 1988).

Table 4–2. Influence of incubation temperature on time and thermal time required for commencement of papermill sludge net N mineralization (adapted from Honeycutt et al., 1988).

Temperature	Net N mineralization requirement	
	Days†	Degree days†
°C		
5	—	—
10	60e	600b
15	46d	684c
20	33c	669c
25	25b	621bc
30	12a	357a

†Values followed by the same letter within a column are not significantly different at $\alpha = 0.05$ with Duncan's multiple range test.

Field microplots using the same rates of papermill sludge were established in the same soil used in the preceding incubation study. Thermal units were calculated by summing the mean daily soil temperatures (7.5-cm depth) above 0°C. Inorganic N concentrations in the field microplots decreased substantially from the first to second sample dates (Fig. 4–8). Although still within the positive range, this decline may be indicative of N immobilization. Considerable levels of net N mineralization began at approximately the third sampling date (Fig. 4–8). Thermal units accumulated during that 29-d period totaled 649 degree days, a value within the 600 to 684 degree day range determined from the 10 to 25°C laboratory treatments (Table 4–2).

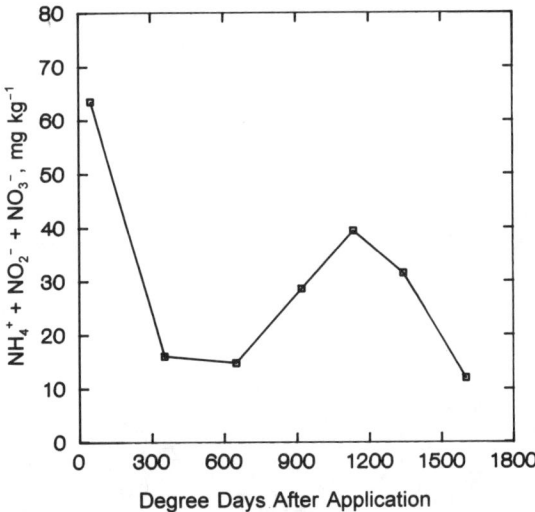

Fig. 4–8. Concentration of $NH_4^+ + NO_2^- + NO_3^-$ in a papermill sludge-amended soil minus that in a control soil related to thermal units in field microplots (Honeycutt et al., 1988).

Crop Residue

Thermal units were next tested for predicting crop residue N mineralization under field climatic conditions. Microplots were established on three dates at 28-d intervals to provide a range in soil temperature, time, and soil water content to rigorously test the thermal unit approach (Honeycutt & Potaro, 1990). Ground (1 mm) corn (cv. King 1113), lupine (*Lupinus albus* L. cv. Ultra), and potato (*Solanum tuberosum* L. cv. Russet burbank) residues were mixed with sieved (2 mm) soil and packed in microplots to a bulk density of 1.0 Mg m^{-3}.

Corn and lupine residues induced a period of net N immobilization, as exemplified by data from the June application (Fig. 4–9). These residues provide a comparison reference point (i.e., where mineralization processes exceed immobilization processes) for assessing the utility of thermal units in predicting N mineralization under field conditions. The immediate net N mineralization from the potato residue precluded using that data for evaluating the thermal unit approach because sampling was restricted to the surface 15 cm (i.e., leaching losses below this depth were not captured).

Net N mineralization from corn and lupine residues was considered to begin on that sample date when extractable (inorganic) N concentrations in these soils were significantly greater ($P < 0.05$) than extractable N in the control soil. Net N mineralization began in both corn and lupine residue treatments in 119, 99, and 317 d, which corresponded to 2346, 1990, and 2360 degree days after residue application in May, June, and July, respectively. Thus a wide range of time (99–317 d) was required for commencement of net N mineralization from the corn and lupine residues. The relative range was greatly reduced upon employing thermal units.

Although the time required for net N mineralization in the May and July plots differed by 198 d, the thermal unit requirement was nearly identical for

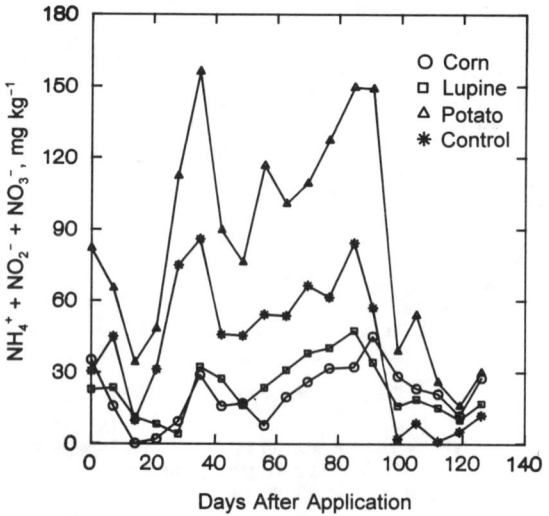

Fig. 4–9. Concentration of $NH_4^+ + NO_2^- + NO_3^-$ in soil microplots treated with four crop residue treatments related to days after residue application (Honeycutt & Potaro, 1990).

these two application months. Of particular significance was the observation that net N mineralization from corn and lupine residues applied in July did not begin until late the following May. This use of thermal units was apparently valid despite rather harsh and variable environmental conditions encountered by the residues and soil.

Factors Affecting Prediction Accuracy

Soil Water

Several factors have been examined to evaluate their impact on N mineralization prediction accuracy with thermal units. Doel et al. (1990) used the same soil and lupine residue previously described by Honeycutt and Potaro (1990) to investigate the influence of soil water on residue N mineralization prediction. Lupine-amended and nonamended soils were incubated for 198 d at factorial combinations of temperature (15, 20, and 25°C) and soil water (–0.30, –0.03, and –0.01 MPa).

Examples depicting the soil water influence on inorganic N dynamics during lupine residue decomposition is provided for the 20°C treatment in Fig. 4–10 to 4–12. All treatments were characterized by an initial period of net N immobilization. Net N mineralization was not observed for the –0.30 MPa treatments by the close of the study (Fig. 4–12).

Commencement of net N mineralization ranged from 103 to 187 d for the remaining treatments (Table 4–3). Thermal units accumulated until initiation of net N mineralization ranged from 2058 to 2814 degree days (Table 4–3) and were generally comparable to slightly higher than those observed (1990–2360 degree days) in the complementary field study (Honeycutt & Potaro, 1990).

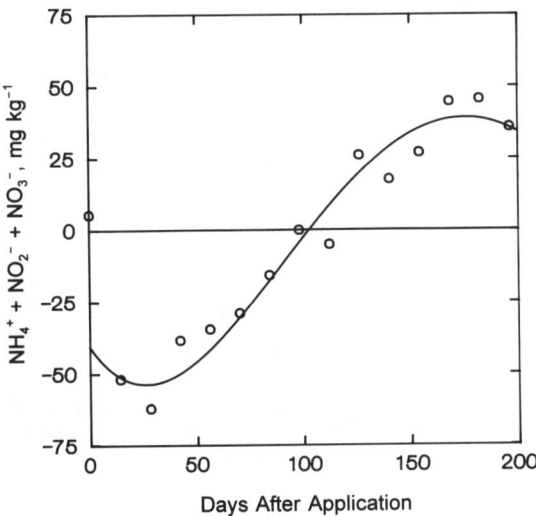

Fig. 4–10. Concentration of $NH_4^+ + NO_2^- + NO_3^-$ in a lupine residue-amended soil minus that in a control soil incubated at 20°C and –0.01 MPa soil water potential (Doel et al., 1990).

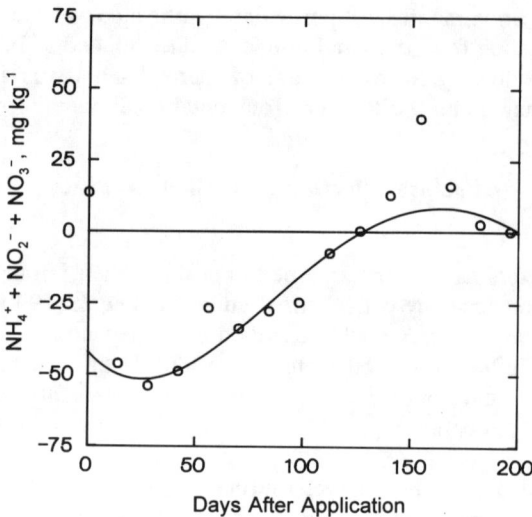

Fig. 4–11. Concentration of $NH_4^+ + NO_2^- + NO_3^-$ in a lupine residue-amended soil minus that in a control soil incubated at 20°C and –0.03 MPa soil water potential (Doel et al., 1990).

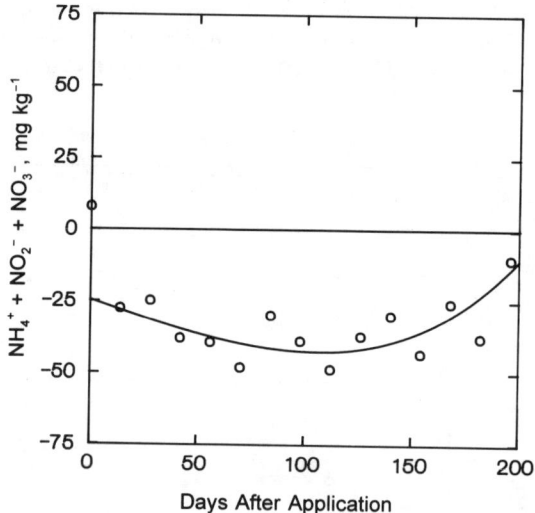

Fig. 4–12. Concentration of $NH_4^+ + NO_2^- + NO_3^-$ in a lupine residue-amended soil minus that in a control soil incubated at 20°C and –0.30 MPa soil water potential (Doel et al., 1990).

Table 4–3. Influence of soil temperature and soil water on time[†] and thermal time[†] required for commencement of lupine residue net N mineralization (from Doel et al., 1990).

	Soil water potential (MPa)			
	−0.03		−0.01	
Temperature	Days[‡]	Degree days[§]	Days[‡]	Degree days[§]
°C				
15	187 (6)a	2802 (94)A	168 (6)a	2514 (97)AB
20	135 (15)c	2696 (135)A	103 (9)b	2058 (175)B
25	113 (11)cb	2814 (175)A	104 (8)b	2598 (192)AB

[†] Mean (standard deviation).
[‡] Values followed by the same lower case letter are not significantly different at $\alpha = 0.05$ with Tukey's HSD range test.
[§] Values followed by the same upper case letter are not significantly different at $\alpha = 0.05$ with Tukey's HSD range test.

Wetting–drying cycles encountered under field conditions may perhaps contribute to this difference.

Thermal units did adequately predict commencement of net N mineralization for five of the six temperature by water treatment combinations (Table 4–3). Absence of net N mineralization at all −0.30 MPa treatments, however, indicates a soil water factor should be considered for incorporation into thermal unit predictions, especially under nonirrigated conditions.

Soil Texture, Climate, and pH

Data from three countries, seven states, and 21 soil series were compiled from the literature to determine the universality of NO_3^- formation vs. thermal unit relations across a range of unamended soils (Honeycutt et al., 1991). Graphical comparisons showed a wide range in NO_3^- concentration when plotted against both time (Fig. 4–13) and thermal time (Fig. 4–14).

Selected chemical, physical, and climatic data characterizing each site were then compiled (Table 4–4) and included as covariates for predicting NO_3^- production. Climatic data were limited to those reported in U.S. Department of Agriculture (1941). Stepwise regression analysis of all unamended soil data resulted in the following equation with $R^2 = 0.937$:

$$NO_3^- = 784.558 - 1.271 \text{ (MAP)} - 93.057 \text{ (pH)} + 0.005 \text{ (DD} \times \text{pH)} + 0.061 \text{ (DD} \times H_2O) \qquad [1]$$

where NO_3^- = NO_3^-–N concentration, mg kg^{-1}
MAP = mean annual precipitation, cm
pH = soil pH
DD = degree days, °C
H_2O = soil water content, MPa

The above model, however, was based on only 240 out of 621 cases due to the absence of data for organic C, total N, and soil water content in many of the

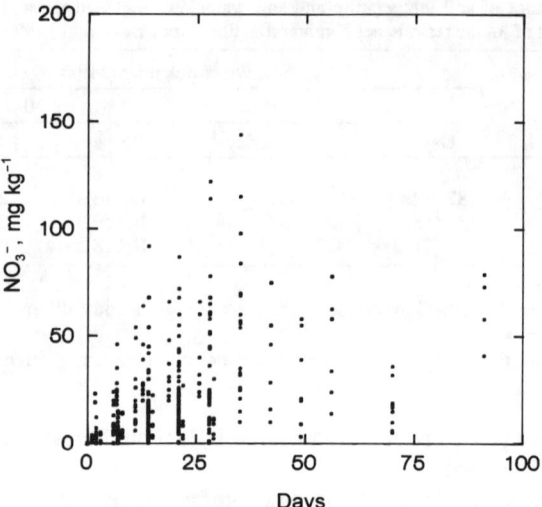

Fig. 4–13. Nitrate concentration from unamended soils related to time (data from studies in Table 4–4).

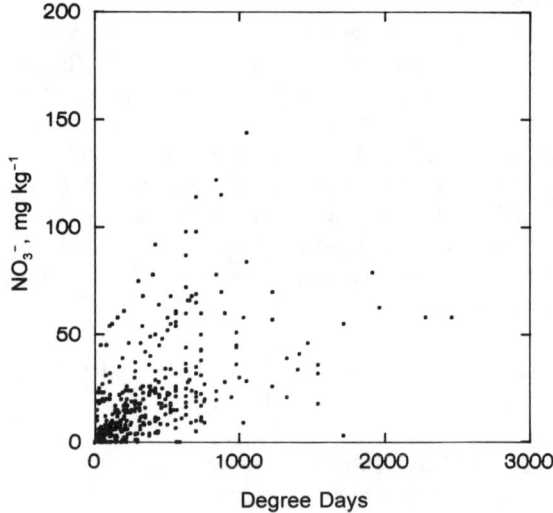

Fig. 4–14. Nitrate concentration from unamended soils related to thermal units (data from studies in Table 4–4).

Table 4-4. Selected chemical, physical, and climatic properties of soils used to assess NO_3^- formation in relation to thermal units (adapted from Honeycutt et al., 1991.)

| Reference | Data file | Soil Classification | | | Sand† | Clay† | Organic C | Total N | Soil water potential | pH | Mean January temp.‡ | Mean July temp.‡ | MAP‡ |
		Series	Family										
					%	%	g kg⁻¹		MPa		°C	°C	cm
Doel et al. (1990)	1	unnamed l	coarse-loamy, mixed, frigid, Typic Haplorthod		42	18	22.0	2.6	−0.30	6.0	−5.0	21.6	100.3
	2	unnamed l	coarse-loamy, mixed, frigid, Typic Haplorthod		42	18	22.0	2.6	−0.03	6.0	−5.0	21.6	100.3
	3	unnamed l	coarse-loamy, mixed, frigid, Typic Haplorthod		42	18	22.0	2.6	−0.01	6.0	−5.0	21.6	100.3
Frederick (1956)	1	Genesee sil	fine-loamy, mixed, nonacid, mesic Typic Udifluvent		20	15	—	—	—	7.7	−2.5	24.5	96.5
	2	Chalmers sicl	fine-silty, mixed, mesic Typic Haplaquoll		10	34	—	—	—	6.2	−2.5	24.5	96.5
	3	Clermont sil	fine-silty, mixed, mesic Typic Glossaqualf		20	15	—	—	—	5.0	−2.5	24.5	96.5
Justice & Smith (1962)	1	Millville l	coarse-silty, carbonatic, mesic Typic Rendoll		41	18	—	1.3	−0.03	7.8	−4.0	23.0	41.9
	2	Millville l	coarse-silty, carbonatic, mesic Typic Rendoll		41	18	—	1.3	−0.10	7.8	−4.0	23.0	41.9
	3	Millville l	coarse-silty, carbonatic, mesic Typic Rendoll		41	18	—	1.3	−1.00	7.8	−4.0	23.0	41.9
Kowalenko & Cameron (1976)§	1	? cl	Humaquepts?		32	34	30.9	2.1	−0.70	6.3	−10.6	21.0	137.0
	2	? cl	Humaquepts?		32	34	30.9	2.1	−0.10	6.3	−10.6	21.0	137.0
	3	? cl	Humaquepts?		32	34	30.9	2.1	−0.012	6.3	−10.6	21.0	137.0
Myers (1975)¶	1	Tindall cl	?		32	34	—	0.45	—	—	22.0	11.6	120.5
Panganiban (1925)	1	Dunkirk sicl	fine-silty, mixed, mesic Glossoboric Hapludalf		10	34	—	1.5	—	6.3	−5.0	21.5	104.6
	2	Clyde cl	fine-loamy, mixed, mesic Typic Haplaquoll		32	34	—	2.7	—	7.0	−5.0	21.5	104.6
	3	Dunkirk fsl	fine-silty, mixed, mesic Glossoboric Hapludalf		65	10	—	0.8	—	6.5	−5.0	21.5	104.6
Parker & Larson (1962)	1	Webster sil	fine-loamy, mixed, mesic Typic Haplaquoll		20	15	—	—	—	7.1	−6.7	24.0	77.4
	2	Marshall sil	fine-silty, mixed, mesic Typic Hapludoll		20	15	—	—	—	6.3	−6.7	24.0	77.4
	3	Colo sicl	fine-silty, mixed, mesic Cumulic Haplaquoll		10	34	—	—	—	6.3	−6.7	24.0	77.4
	4	Monona sil	fine-silty, mixed, mesic Typic Hapludoll		20	15	—	—	—	6.3	−6.7	24.0	77.4
	5	Grundy sil	fine, montmorillonitic, mesic Aquic Argiudoll		20	15	—	—	—	5.9	−6.7	24.0	77.4
Russell et al. (1925)	1	Carrington cl	inactive soil family		35	21	27.4	2.2	—	—	−4.0	25.6	69.9
	2	Colby vfsl	fine-silty, mixed (calcareous), mesic Ustic Torriorthent		66	16	12.7	1.2	—	—	−3.9	24.0	46.2
Tyler et al. (1959)	1	Hanford sl	coarse-loamy, mixed, nonacid, thermic Typic Xerorthent		65	10	—	—	—	6.4	7.2	24.0	41.7
	2	Sacramento c	very-fine, montmorillonitic, thermic Vertic Haplaquoll		20	60	—	—	—	5.6	7.2	24.0	41.7
	3	Yolo l	fine-silty, mixed, nonacid, thermic Typic Xerorthent		41	18	—	—	—	7.6	7.2	24.0	41.7
	4	Salinas c	fine-loamy, mixed, thermic Pachic Haploxeroll		20	60	—	—	—	8.1	7.2	24.0	41.7

†Sand and clay contents were reported by Russell et al. (1925) and Doel et al. (1990). All others were estimated from given soil textures as the mean content of that textural classification (Soil Survey Staff, 1975).
‡Provided by U.S. Department of Agriculture (1941); MAP = mean annual precipitation.
§Climatic data are for Montreal, Canada.
¶Climatic data are for Sydney, Australia.

studies (Table 4–4). Omission of these covariates from the stepwise analysis yielded the following equation with $n = 542$ and $R^2 = 0.777$:

$$NO_3^- = 770.610 + 1.003 \text{ (SILT)} - 13.944 \text{ (pH)} + 7.714 \text{ (MAT)} \\ - 32.528 \text{ (JULY TEMP)} - 0.005 \text{ (DD} \times \text{JAN TEMP)} - 2.356 \text{ (PT)} \\ + 0.015 \text{ (DAY} \times \text{CLAY)} \quad [2]$$

where   SILT = silt content, %
MAT = mean annual temperature, °C, estimated by the mean of JULY TEMP and JAN TEMP
JULY TEMP = mean July temperature, °C
JAN TEMP = mean January temperature, °C
PT = MAP/MAT
CLAY = clay content, %

Identification of these variables suggests the importance of soil texture, pH, and climate on the ammonifying and nitrifying populations, as supported by Morrill and Dawson (1967), Dancer et al. (1973), Malhi and McGill (1982), and Schimel et al. (1985).

The influence of soil texture is perhaps a surrogate for soil water as it relates to microbial activity and plant productivity (i.e., soil organic matter inputs, pool sizes, and pool labilities). Identification of climate may likewise reflect such soil organic matter inputs and properties. Climatic variables, however, may also reflect the importance of climate on microbial evolution. The later may express itself through artifact production when these soils are subjected to climatic regimes not normally encountered under field conditions (e.g., the 30°C data in Fig. 4–3 and 4–7). Again, this reasoning leads to cautious use of the 35°C treatment prescribed for all soils in determining N mineralization potential (Stanford & Smith, 1972).

Crop Residue Loading Rate and Quality

Both significant (Broadbent & Bartholomew, 1948; Hallam & Bartholomew, 1953; Stott et al., 1990) and nonsignificant (Jenkinson, 1971, 1977; Douglas, 1989) relationships between crop residue loading rate and its decomposition have been reported. The influence of crop residue quality (i.e., chemical composition) on N mineralization is also widely recognized, although identification of parameters important for characterizing quality has been inconsistent (Iritani & Arnold, 1960; Allison & Klein, 1962; Herman et al., 1977; Frankenberger & Abdelmagid, 1985; Moorhead et al., 1987; Janzen & Kucey, 1988).

The independent and interactive effects of residue quantity and quality on predicting N mineralization with thermal units was examined by Honeycutt et al. (1993). Plant material of different qualities was obtained by field-planting hairy vetch (*Vicia villosa* Roth) in May 1988 and sampling the following September 1988 (Q1) and April 1989 (Q2). Vetch biomass averaged 5212 and 2955 kg ha^{-1} (dry weight basis) for the Q1 and Q2 sample dates, respectively. Vetch loading rates in the laboratory incubation study corresponded to approximately 0, 2150, 4300, and 6450 kg ha^{-1}. A three temperature (15, 20, and 25°C) × two residue quality × four loading rate factorial design was employed.

Extractable (mineralized) N in the residue treated soils was related to residue loading rate over both time and thermal units (Table 4–5). Nitrogen mineralization was also related to significant loading rate × day and loading rate × degree day interactions (Table 4–5, Fig. 4–15 and 4–16).

Differing relationships between thermal units and residue N mineralization among loading rates would translate into a considerable effort needed to develop and transfer these relationships to a pragmatic, field level. Expressing vetch N mineralization as a percentage of the total N added with each loading rate, however, allows a very different view of these data. The independent effect of loading rate and its interactive effect with either day or degree day did not influence the percentage of added N mineralized (Table 4–5). Rather, similar relationships between thermal units and percentage of added N mineralized were observed across loading rates for a given residue quality (Fig. 4–17 and 4–18). Therefore, practical application of thermal units for predicting vetch residue N mineralization does not appear limited by differential residue loading rates representative of those encountered under field conditions. Knowledge of a particular residue's total N content and the relationship between percentage of that N mineralized and thermal units should allow prediction of available N level regardless of residue loading rate.

Table 4–5. Summary of analysis of variance for extractable N and percentage of added N mineralized with each modeled day and degree day† (from Honeycutt et al., 1993).

Effect‡	Extractable N vs.		Percentage of added N mineralized vs.	
	Day	Degree day	Day	Degree Day
T	***	—	**	—
Q	***	***	***	***
L	***	***	NS	NS
T × Q	NS	—	NS	—
T × L	NS	—	*	—
Q × L	***	***	**	NS
T × Q × L	NS	—	*	—
D	***	—	***	—
D × T	***	—	*	—
D × Q	***	—	***	—
D × L	***	—	NS	—
D × T × Q	NS	—	NS	—
D × T × L	NS	—	NS	—
D × T × Q × L	NS	—	NS	—
DD	—	***	—	***
DD × Q	—	***	—	***
DD × L	—	***	—	NS
DD × Q × L	—	***	—	NS

*, **, *** Significant at $\alpha = 0.05$, 0.01, and 0.001, respectively; NS = not significant at $\alpha = 0.05$.
†Analysis of variance included all four loading rates for extractable N but only three loading rates for percentage of added N mineralized because the latter involved subtraction of extractable N in unamended soils from that in residue-amended soils.
‡T = temperature; Q = quality; L = loading rate.

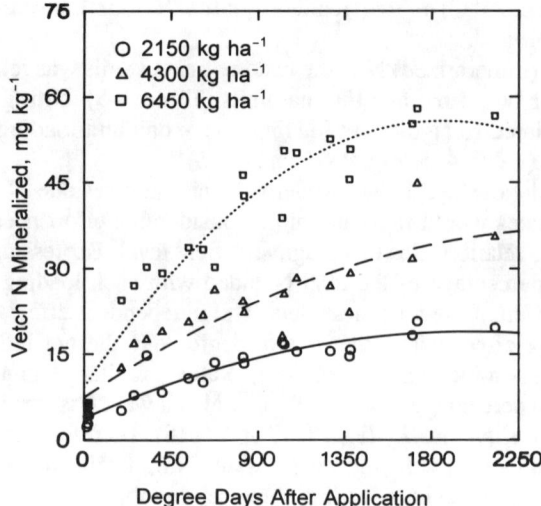

Fig. 4–15. Relationships between thermal units and vetch N mineralization for three loading rates of residue Quality 1 (Q1) (Honeycutt et al., 1993).

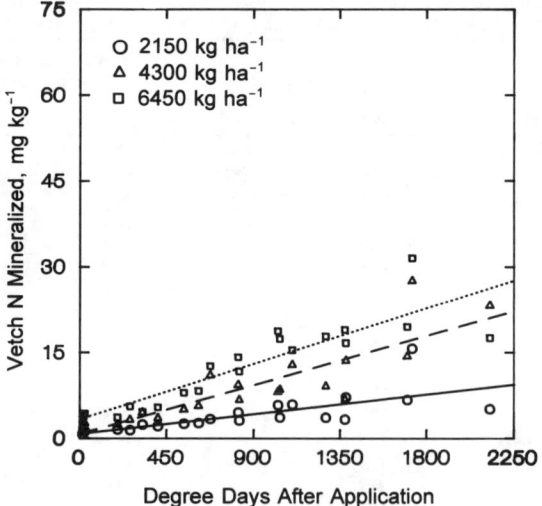

Fig. 4–16. Relationships between thermal units and vetch N mineralization for three loading rates of residue Quality 2 (Q2) (Honeycutt et al., 1993).

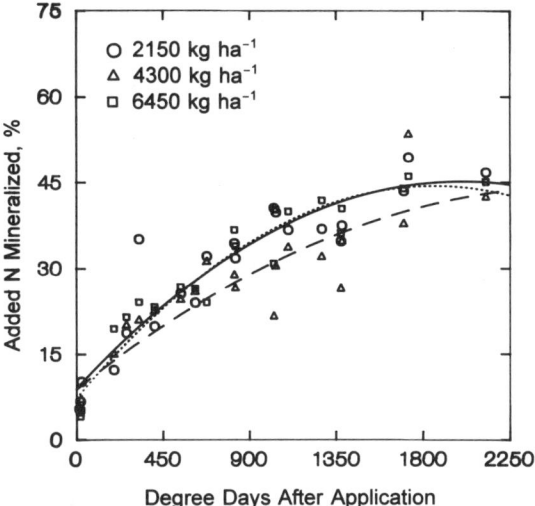

Fig. 4–17. Relationships between thermal units and percentage of added N mineralized for three loading rates of residue Quality 1 (Q1) (Honeycutt et al., 1993).

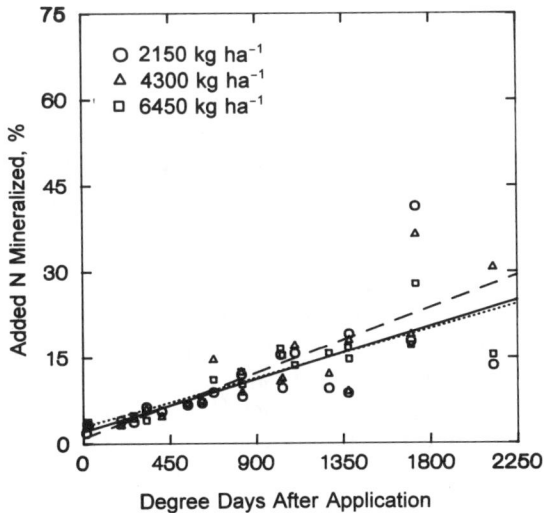

Fig. 4–18. Relationships between thermal units and percentage of added N mineralized for three loading rates of residue Quality 2 (Q2) (Honeycutt et al., 1993).

Table 4-6. Selected chemical characteristics† of vetch residues examined by Honeycutt et al. (1993).

Characteristic	Q1‡	Q2‡
Total C, g kg^{-1}	456.0 (1.9)	430.2 (2.7)
Total H, g kg^{-1}	63.3 (0.3)	58.9 (1.1)
Total N, g kg^{-1}	39.6 (0.2)	35.7 (0.2)
Total C/Total H	7.21 (0.03)	7.30 (0.11)
Total C/Total N	11.52 (0.04)	12.04 (0.03)
Hemicellulose, g kg^{-1}	138.2 (16.7)	236.4 (23.3)
Cellulose, g kg^{-1}	201.1 (2.9)	191.8 (10.0)
Lignin, g kg^{-1}	72.4 (2.9)	142.4 (8.8)

†Mean (standard deviation) of five to six determinations. All values are expressed on an oven-dry (65°C) basis.
‡Q1, residue Quality 1 (harvested September, 1988); Q2, residue Quality 2 (Harvested April, 1989).

Residue quality exhibited a marked effect on both extractable N and the percentage of added N mineralized (Table 4-5). Approximately 45% of residue N added was predicted to be mineralized 2250 degree days after application of Q1 (Fig. 4-17). This contrasts with ~25 to 30% of added N mineralized 2250 degree days after application of Q2 (Fig. 4-18).

Thermal units were related to both expressions of residue N mineralization in a quadratic fashion ($P < 0.001$) for Q1 (Fig. 4-15 and 4-17). Thermal units, however, were linearly related to residue N mineralization for Q2, with coefficients for the quadratic terms determined to be nonsignificant ($\alpha = 0.05$) (Fig. 4-16 and 4-18). Therefore, not only did residue quality affect the percentage of added N mineralized, but it also affected net N mineralization rate and overall shape of the response (Fig. 4-17 and 4-18). These data indicate that residue quality must be considered when developing crop residue N mineralization × thermal unit relationships for predictive purposes. The fact that this association is demonstrated using different developmental stages of the same plant species substantiates reports that factors other than plant species should be used in defining plant residue quality (Müller et al., 1988).

Examination of the chemical composition characterizing each residue shows relatively similar total N contents and C/N ratios (Table 4-6). Both lignin and hemicellulose contents increased almost two-fold from the fall (Q1) to the spring (Q2) residue collection dates (Table 4-6). The significant relationship between residue quality and vetch N mineralization may be due to the differential concentrations of lignin or hemicellulose rather than N content or C/N ratio.

RECENT APPLICATIONS OF THERMAL UNITS

Thermal units have recently been applied to other situations. Wolf and Rogowski (1991) used estimates of soil temperature and heat capacity to investigate spatial distribution of soil heat flux and thermal units at both farm and watershed scales. This allowed delineation of areas with differing C and N mineralization rates at both scales. Wolf and Rogowski (1991) proposed combining soil thermal unit predictions with manure and sludge application practices to minimize groundwater pollution potential.

Thermal units have recently been shown to be a powerful predictor of crop residue decomposition for a wide array of residues and climates when combined with information on residue placement and initial N content (Douglas & Rickman, 1992). Thermal units successfully predicted decomposition of soft white and hard red spring and winter wheat (*Triticum aestivum* L.), durum wheat (*T. durum* Desf.), spring and winter barley, triticale (x *Triticosecale* Wittm.), corn, and soybean [*Glycine max* (L.) Merr.] from Alaska, Idaho, Indiana, Missouri, Oregon, Texas, and Washington (Douglas & Rickman, 1992). Successful prediction of residue decomposition has important applications for soil conservation because crop residues significantly impact soil erosion.

LINKING NITROGEN MINERALIZATION WITH PLANT NITROGEN DEMAND

Another important area of thermal unit application is nutrient management. Thermal units provide a common, biologically meaningful unit to compare N mineralization predictions with plant N demand. Relevant budget approaches to N management have been previously based on total crop N uptake or total fertilizer N requirements (Parr, 1973). Consequently, the dynamics of plant N demand in relation to the dynamics of soil N availability were not considered with sufficient detail (i.e., an adequate time step) for assessing these relationships during the course of a growing season.

Several reports have called for increased N use efficiency through improved synchrony between soil N availability and crop N demand (Pierce & Rice, 1988; Honeycutt & Potaro, 1990). A formal, yet nonvalidated approach for addressing these relationships was recently proposed (Honeycutt et al., 1994). The proposition is founded on mass balance principles (Parr, 1973; Meisinger, 1984) and evidence previously presented that thermal units can be used to describe plant growth (i.e., N uptake) and N mineralization from soils and soil amendments. Briefly, plant-available N is determined from studies of N mineralization from soil organic matter and other organic N sources. Functions describing N use efficiency dynamics are used to predict N uptake from these sources. Nitrogen uptake by a crop when grown at its optimal N fertilizer rate is then compared with predicted N uptake from mineralized organic sources. This allows calculation of additional N required by the plant. The following example is provided using potato (cv. Norwis) grown continuously and in rotation with hairy vetch. This example is taken from Honeycutt et al. (1994).

Steps of the Approach

Nitrogen Mineralization from Soil Organic Matter

Total soil N (N_{ts}) and bulk density are first determined using appropriate methods (Bremner & Mulvaney, 1982; Blake & Hartge, 1986). Initial soil inorganic N (N_i) and N mineralization from soil organic matter (N_{is}) vs. thermal unit relations are determined by laboratory incubation (Honeycutt et al., 1988). The latter relationship is expressed as a percentage of N_{ts} mineralized.

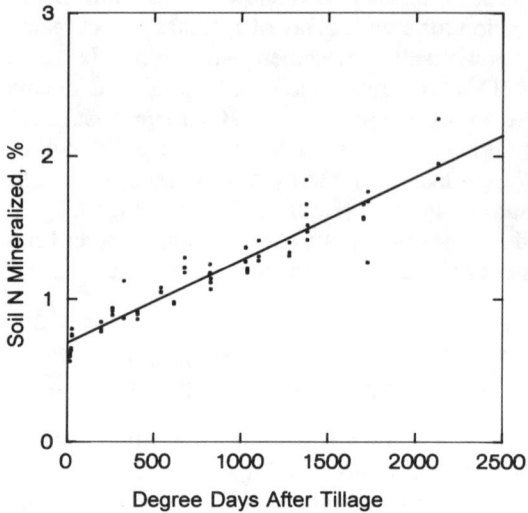

Fig. 4–19. Percentage of soil N mineralized in relation to thermal units after incorporation (Honeycutt et al., 1994).

Incubation commencement may be considered to represent the first day of tillage. For example, the surface 15 cm of an agricultural soil classified as a coarse-loamy, mixed, frigid Typic Haplorthod (Soil Survey Staff, 1975) was sampled and incubated at known temperatures in the laboratory (Fig. 4–19) (Honeycutt et al., 1993). The following function describes the relationship shown in Fig. 4–19.

$$PN_{is} = 0.693690 + 0.000580 \, (DDAT) \qquad [3]$$

where $PN_{is} = N_{ts}$ mineralized, %
DDAT = soil degree days after tillage, °C

Soil degree days are calculated by summing the average daily soil temperature above 0 °C (Honeycutt et al., 1988; Honeycutt & Potaro, 1990). For constant temperature laboratory incubations, average daily soil temperature is assumed equal to incubator air temperature. Under field conditions, soil temperatures are measured at the 7.5-cm depth with thermocouples or thermistors interfaced with a datalogger, or alternatively, with a maximum–minimum thermometer that is read daily.

Total soil N is expressed on a weight per area basis for that soil's bulk density and depth of sampling or tillage. A soil with a bulk density of 1.0 Mg m^{-3}, N_{ts} = 2.6 g kg^{-1}, and tilled to a 15.24-cm depth contains: (1 524 000 kg soil ha^{-1})(0.0026) = 3962.4 kg N ha^{-1} in the 15.24-cm thickness. Equation [3] is multiplied by this value to determine plant-available N mineralized from soil organic matter (N_{is}) expressed in kilograms per hectare.

$$N_{is} = 27.486773 + 0.022982 \, (DDAT) \qquad [4]$$

Mineralization of Organic Nitrogen Additions

Nitrogen mineralization from additional sources such as crop residue, animal manure, and sludge should also be determined in relation to thermal units if these are pertinent to the cropping system. Nitrogen mineralization from vetch residue is considered in the following example.

Vetch N mineralization vs. thermal unit relations were previously determined to be independent of the quantity of residue added (Honeycutt et al., 1993). These data were therefore pooled, and the relationship determined in Eq. [5] and depicted in Fig. 4-20 can be used for different residue quantities provided residue qualities are similar. (Pooling all data across residue loading rates of a given residue quality probably resulted in the wide scatter observed in Fig. 4-20).

$$PN_{ir} = 2.006272 + 0.010785 \, (DDAT) \quad [5]$$

where PN_{ir} = residue N mineralized, %

For a vetch residue biomass of 3266 kg ha^{-1} with a N concentration of 35.7 g kg^{-1}, 116.596 kg N ha^{-1} is added to the soil as vetch residue. Plant-available N from the residue (N_{ir}) as related to DDAT may then be determined from Eq. [5] and expressed in kilograms per hectare:

$$N_{ir} = 2.339233 + 0.012575 \, (DDAT) \quad [6]$$

Total Plant-Available Nitrogen

Total plant-available N (N_a) as related to thermal units may be determined by simply adding the function describing N_{is} with those functions describing N

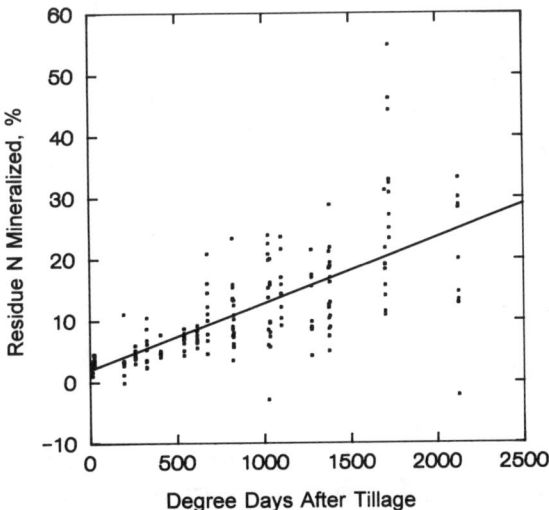

Fig. 4-20. Percentage of hairy vetch residue N mineralized in relation to thermal units after incorporation (Honeycutt et al., 1994).

mineralization from all organic additions. In this example, Eq. [4] and Eq. [6] are added to give:

$$N_a = 29.826006 + 0.035557 \text{ (DDAT)} \quad [7]$$

Thus far, N_a dynamics have been related to DDAT. This is because tillage can dramatically impact decomposition and N mineralization of soil organic matter and supplemental organic N. Field application of the proposed approach must therefore include monitoring soil temperature commencing with the first day of tillage.

Historical Nitrogen Use Efficiency

Plant uptake from each N pool has an associated N use efficiency (Pierce & Rice, 1988). To estimate that efficiency during the course of a growing season, crop N uptake (N_u) can be related to N_a over thermal time. Continuing with the same example, N_u by potato (shoots plus roots plus tubers) following hairy vetch with no N fertilizer in 1989 was related to soil degree days after planting (DDAP) (Fig. 4–21) by the following:

$$N_u = 0.000000000 - 0.050089812 \text{ (DDAP)} + 0.000131785 \text{ (DDAP)}^2 \\ - 0.000000044 \text{ (DDAP)}^3 \quad [8]$$

The constant in Eq. [8] was forced through zero to set $N_u = 0$ on the day of planting.

Predicted N_a related to DDAT may be determined by solving Eq. [4] and Eq. [6] considering bulk density, total soil organic N, vetch residue biomass, and vetch residue N concentration measured in 1989. Plant-available N can then be related to DDAP by solving for the number of degree days accumulated between tillage and planting. Thus a new y-intercept is determined allowing plant-available N and plant uptake of N (expressed in the same units) to be compared over thermal time (Fig. 4–21).

A historical N use efficiency (NUE) for N mineralized from soil organic matter and vetch residue may be calculated by dividing Eq. [8] by the N_a vs. DDAP function for 1989. In this example, historical NUE (expressed as a percentage) is:

$$NUE = (100)[0.000000000 - 0.050089812(DDAP) + 0.000131785(DDAP)$$

$$-0.000000044(DDAP)^3 \,]/[38.698342 + 0.034261(DDAP)], \quad [9]$$

which reduces to:

$$NUE = (100)\frac{-DDAP\left(44(DDAP)^2 - 131785(DDAP) + 50089812\right)}{1000(34261(DDAP) + 38698342)} \quad [10]$$

Equation [10] is depicted in Fig. 4–22. It is common to consider NUE as a single value, e.g., plant recovery of a single N application. It is clear from Fig. 4–22, however, that N recovery from soil organic matter and crop residue is more accurately

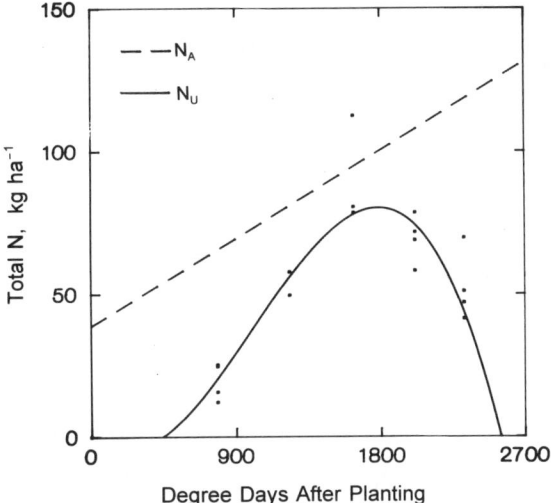

Fig. 4–21. Total N uptake by potatoes following hairy vetch in 1989 (solid line) and N predicted to have mineralized from vetch plus soil organic matter in 1989 (dashed line) in relation to thermal units after planting (Honeycutt et al., 1994).

described as a dynamic efficiency, one that reflects changing plant N availability with changing plant N demand. Estimates of N_a as shown here reflect net N mineralization, but not N losses associated with leaching and denitrification. Measurements of N_u do indirectly reflect those losses. Incorporating leaching and denitrification losses from the available N pool, however, should improve NUE function accuracy.

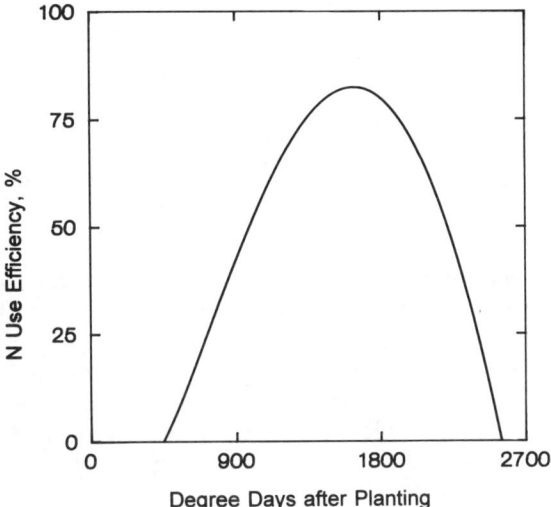

Fig. 4–22. Nitrogen use efficiency in 1989 related to thermal units after planting (Honeycutt et al., 1994).

Predicted Uptake of Plant-Available Nitrogen

Uptake of N_a by the current crop may be predicted ($\hat{N}_u$) by multiplying the historical NUE function by the function describing N_a for the current year's N_{ts}, bulk density, crop residue biomass, and crop residue N concentration.

$$\hat{N}_u = \text{NUE}(N_a) \qquad [11]$$

In this example, 116 degree days accumulated between tillage and planting. Equation [12] is then determined by solving Eq. [7] for DDAT = 116.

$$N_a = 33.950618 + 0.035557 \, (\text{DDAP}) \qquad [12]$$

and

$$\hat{N}_u = \left[\frac{-DDAP\left(44(DDAP)^2 - 131785(DDAP) + 50089812\right)}{1000(34261(DDAP) + 38698342)} \right] \qquad [13]$$

$$\times (33.950618 + 0.035557(DDAP)),$$

which reduces to:

$$\hat{N}_u = -[DDAP(35557(DDAP) + 33950618)(44(DDAP)^2 - 131785(DDAP)$$

$$+ 50089812)] / [1000000000(34261(DDAP) + 38698342)].$$

Additional Nitrogen Requirement

A functional relationship between DDAP and crop N uptake at the economically optimum or recommended fertilizer N rate (N_{ou}) can be considered a targeted function to attain for optimum yield. This relationship (Eq. [15]) is shown by the solid line in Fig. 4–23 for continuous potato grown in 1989 at the recommended N fertilizer rate of 179 kg ha^{-1}.

$$N_{ou} = 0.000000000 - 0.040213382(\text{DDAP}) + 0.000165991(\text{DDAP})^2$$
$$- 0.000000058(\text{DDAP})^3 \qquad [15]$$

Predicted uptake of plant-available N from soil organic matter and crop residue (Eq. [14]) is shown as a dashed line in Fig. 4–23. The area between curves represents the predicted plant N deficit for the current year if N mineralized from soil organic matter and vetch residue are the only sources of available N. Subtracting predicted N uptake from vetch and soil organic matter from that N uptake needed for optimum yield provides the additional plant N requirement (N_{ADD}) in relation to thermal units (Eq. [16] and Fig. 4–24).

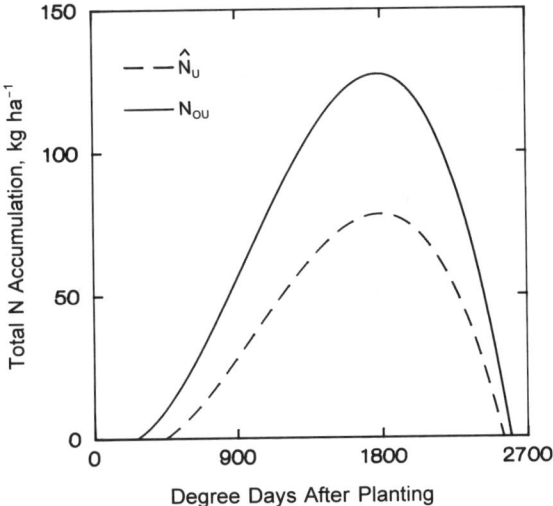

Fig. 4–23. Total N uptake by potato required for optimum yield in 1989 (solid line) and predicted N uptake by potato following hairy vetch for 1992 (dashed line) in relation to thermal units after planting (Honeycutt et al., 1994).

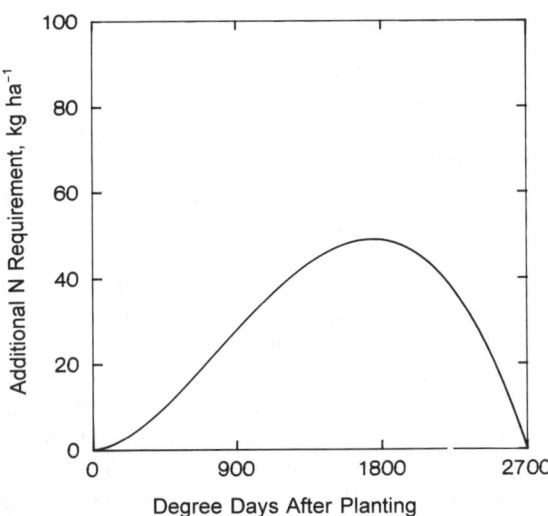

Fig. 4–24. Additional N predicted to be required by potato in 1992 as related to thermal units after planting (Honeycutt et al., 1994).

$$N_{ADD} = N_{ou} - N_u$$

$$= 53.6016(DDAP)^5 - 48819(DDAP)^4 - (2.15796 \times 10^8)(DDAP)^3 - (4.59564 \times 10^{10})(DDAP)^2 + (6.58096 \times 10^{13})(DDAP) \qquad [16]$$

The previous function provides the grower with information on quantity of additional N to apply and when, in thermal time, it should be applied. Both factors can dramatically affect N fertilizer use efficiency (Olson & Kurtz, 1982; Meisinger, 1984). It is important to recognize that N_{ADD} represents the additional plant N requirement, which would equal the amount of N to apply only in the case of 100% uptake efficiency. Consequently, the ability to provide the plant with N_{ADD} introduces another efficiency term. For example, assume a grower desires to apply N fertilizer at planting to meet the potato crop's N requirement up to 900 DDAP (at which time additional N should be applied according to Fig. 4–24). Solving Eq. [16] shows the crop requires ~28 kg N ha^{-1} by 900 DDAP. The grower must then assume a certain uptake efficiency for that application. If 50% recovery is assumed, then 56 kg N ha^{-1} should be applied at planting with additional N applied 900 DDAP. The grower must also assume a certain recovery of the second N application. These efficiency terms can, however, be experimentally determined to serve as future guidelines.

Simply comparing N_a (Fig. 4–21) with N_{ou} (Fig. 4–23) would indicate N from vetch and soil organic matter could meet the crop's optimal N uptake requirement until ~1080 DDAP. Indeed, the term *synchrony* would imply a similar interpretation. Such an approach, however, does not consider NUE dynamics throughout the growing season and would consequently lead to an incorrect N management recommendation. Thermal units provide a common, biologically meaningful unit of measure to calculate NUE dynamics.

Assumptions and Limitations

Several assumptions and limitations must be considered when implementing the proposed approach. One assumption is that historical NUE functions predict NUE dynamics in subsequent growing seasons. Another is that N mineralization in the field proceeds similarly to that determined in the laboratory. Studies have indicated field N mineralization can be predicted from laboratory incubations by employing thermal units (Honeycutt et al., 1988; Honeycutt & Potaro, 1990; Doel et al., 1990). Long periods of dry soil conditions, however, may prevent this information transfer due to reduced microbial activity at low moisture contents (Doel et al., 1990). Both of the preceding assumptions may be more limiting in dryland rather than irrigated crop production.

Functions describing N_a also do not reflect leaching or denitrification losses. Historical plant N uptake relations, however, do indirectly reflect those losses (Fig. 4–21). Models of leaching and denitrification could conceivably be interfaced with the proposed approach. The thermal control on denitrification could make it particularly amenable to thermal unit prediction, especially if temperature interactions with C and O supply are considered (Firestone, 1982).

It is also assumed that application of N fertilizer will not dramatically affect N mineralization from soil organic matter and supplemental organic N. Such priming effects can be determined and incorporated. Lastly, miscellaneous inputs (e.g., precipitation) and outputs (e.g., erosion) to the available N pool are considered minimal and offsetting.

One limitation is that N mineralization and nitrification relations with thermal units are apparently soil-specific, indicating these relationships should be determined for soils with contrasting textures, pH, and climates (Honeycutt et al., 1991). The same can be stated for differing crop species, cultivars, and management schemes (e.g., till vs. no-till).

Research is currently underway to validate the proposed approach under field conditions, substitute plant growth functions for several equations presented, and incorporate algorithms to reflect leaching and denitrification losses from the N_a pool.

CONCLUSIONS

Several studies have demonstrated the utility of thermal units for describing the combined temporal and thermal controls on N mineralization from soil organic matter, crop residues, and papermill sludge. Limitations of the approach have also been identified, particularly those related to soil water, soil texture, soil pH, climate, and crop residue quality. Within these constraints, thermal units are offered to link N mineralization with plant N demand. A function describing N use efficiency dynamics over thermal time can then be calculated to predict N uptake from mineralized organic sources in subsequent growing seasons. Additional N is prescribed by comparing this prediction with N uptake by a crop grown at a fertilizer N rate exhibiting optimum yield. This proposition is largely unvalidated at the time of this writing, and additional validation efforts are invited.

REFERENCES

Allison, F.E., and C.J. Klein. 1962. Rates of immobilization and release of nitrogen following additions of carbonaceous materials and nitrogen to soils. Soil Sci. 92:383–386.

Andrén, O., and K. Paustian. 1987. Barley straw decomposition in the field: A comparison of models. Ecol. 68:1190–1200.

Atlas, R.M. 1988. Microbiology: Fundamentals and applications. 2nd ed. Macmillan, New York.

Blake, G.R., and K.H. Hartge. 1986. Bulk density. p. 363–375. In A. Klute (ed.) Methods of soil analysis. Part 1. 2nd ed. Agron. Monogr. 9. ASA and SSSA, Madison, WI.

Bremner, J.M., and C.S. Mulvaney. 1982. Nitrogen-total. p. 595–624. In A.L. Page et al. (ed.) Methods of soil analysis. Part 2. 2nd ed. Agron. Monogr. 9. ASA and SSSA, Madison, WI.

Broadbent, F.E., and W.V. Bartholomew. 1948. The effect of quantity of plant material added to soil on its rate of decomposition. Soil Sci. Soc. Am. Proc. 13:271–274.

Cabrera, M.L., and D.E. Kissel. 1988. Evaluation of a method to predict nitrogen mineralized from soil organic matter under field conditions. Soil Sci. Soc. Am. J. 52:1027–1031.

Campbell, C.A., V.O. Biederbeck, and F.G. Warder. 1971. Influence of simulated fall and spring conditions on the soil system: II. Effect on soil nitrogen. Soil Sci. Soc. Am. Proc. 35:480–483.

Cross, H.Z., and M.S. Zuber. 1972. Prediction of flowering dates in maize based on different methods of estimating thermal units. Agron. J. 64:351–355.

Dancer, W.S., L.A. Peterson, and G. Chesters. 1973. Ammonification and nitrification of N as influenced by soil pH and previous N treatments. Soil Sci. Soc. Am. Proc. 37:67–69.

Doel, D.S., C.W. Honeycutt, and W.A. Halteman. 1990. Soil water effects on the use of heat units to predict crop residue carbon and nitrogen mineralization. Biol. Fertil. Soils 10:102–106.

Douglas, C.L., Jr. 1989. Straw loading rate and field placement effects on decomposition of wheat straw with different nitrogen contents. p. 213. *In* Agronomy abstracts. ASA, Madison, WI.

Douglas, C.L., Jr., and R.W. Rickman. 1992. Estimating crop residue decomposition from air temperature, initial nitrogen content, and residue placement. Soil Sci. Soc. Am. J. 56:272–278.

Firestone, M.K. 1982. Biological denitrification. p. 289–326. *In* F.J. Stevenson (ed.) Nitrogen in agricultural soils. Agron. Monogr. 22. ASA, CSSA, and SSSA, Madison, WI.

Frankenberger, W.T., Jr., and H.M. Abdelmagid. 1985. Kinetic parameters of nitrogen mineralization rates of leguminous crops incorporated into soil. Plant Soil 87:257–271.

Frederick, L.R. 1956. The formation of nitrate from ammonium nitrogen in soils: I. Effect of temperature. Soil Sci. Soc. Am. Proc. 20:496–500.

Gilmore, E.C., Jr., and J.S. Rogers. 1958. Heat units as a method of measuring maturity in corn. Agron. J. 50:323–326.

Gilmour, C.M., F.E. Broadbent, and S.M. Beck. 1977. Recycling of carbon and nitrogen through land disposal of various wastes. p. 173–194. *In* L.F. Elliott and F.J. Stevenson (ed.) Soils for management of organic wastes and waste water. ASA, CSSA, and SSSA, Madison, WI.

Griffin, G.F., and A.F. Laine. 1983. Nitrogen mineralization in soils previously amended with organic wastes. Agron. J. 75:124–129.

Hallam, M.J., and W.V. Bartholomew. 1953. Influence of rate of plant residue addition in accelerating the decomposition of soil organic matter. Soil Sci. Soc. Am. Proc. 17:365–368.

Hallberg, G.R. 1987. Agricultural chemicals in ground water: Extent and implications. Am. J. Altern. Agric. 2:3–15.

Herman, W.A., W.B. McGill, and J.F. Dormaar. 1977. Effects of initial chemical composition on decomposition of roots of three grass species. Can. J. Soil Sci. 57:205–215.

Honeycutt, C.W., W.M. Clapham, and S.S. Leach. 1994. A functional approach to efficient N use in crop production. Ecol. Model. 72:51–61.

Honeycutt, C.W., and L.J. Potaro. 1990. Field evaluation of heat units for predicting crop residue carbon and nitrogen mineralization. Plant Soil 125:213–220.

Honeycutt, C.W., L.J. Potaro, K.L. Avila, and W.A. Halteman. 1993. Residue quality, loading rate and soil temperature relations with hairy vetch (*Vicia villosa* Roth) residue carbon, nitrogen and phosphorus mineralization. Biol. Agric. Hortic. 9:181–199.

Honeycutt, C.W., L.J. Potaro, and W.A. Halteman. 1991. Predicting nitrate formation from soil, fertilizer, crop residue, and sludge with thermal units. J. Environ. Qual. 20:850–856.

Honeycutt, C.W., L.M. Zibilske, and W.M. Clapham. 1988. Heat units for describing carbon mineralization and predicting net nitrogen mineralization. Soil Sci. Soc. Am. J. 52:1346–1350.

Iritani, W.M., and C.Y. Arnold. 1960. Nitrogen release of vegetable crop residues during incubation as related to their chemical composition. Soil Sci. 89:74–82.

Janzen, H.H., and R.M.N. Kucey. 1988. C, N, and S mineralization of crop residues as influenced by crop species and nutrient regime. Plant Soil 106:35–41.

Jenkinson, D.S. 1971. Studies on the decomposition of 14-C labelled organic matter in soil. Soil Sci. 111:64–70.

Jenkinson, D.S. 1977. Studies on the decomposition of plant material in soil: IV. The effect of rate of addition. J. Soil Sci. 28:417–423.

Justice, J.K., and R.L. Smith. 1962. Nitrification of ammonium sulfate in a calcareous soil as influenced by combinations of moisture, temperature, and levels of added nitrogen. Soil Sci. Soc. Am. Proc. 26:246–250.

Keeney, D.R. 1982. Nitrogen management for maximum efficiency and minimum pollution. p. 605–649. *In* F.J. Stevenson (ed.) Nitrogen in agricultural soils. Agron. Monogr. 22. ASA, CSSA, and SSSA, Madison, WI.

Klepper. B., R.W. Rickman, and C.M. Peterson. 1982. Quantitative characterization of vegetative development in small cereal grains. Agron. J. 74:789–792.

Kowalenko, C.G., and D.R. Cameron. 1976. Nitrogen transformations in an incubated soil as affected by combinations of moisture content and temperature and adsorption-fixation of ammonium. Can. J. Soil Sci. 56:63–70.

Lehenbauer. P.A. 1914. Growth of maize seedlings in relation to temperature. Phys. Res. 1:247–288.

Malhi, S.S., and W.B. McGill. 1982. Nitrification in three Alberta soils: Effect of temperature, moisture and substrate concentration. Soil Biol. Biochem. 14:393–399.

Marion, G.M., J. Kummerow, and P.C. Miller. 1981. Predicting nitrogen mineralization in chaparral soils. Soil Sci. Soc. Am. J. 45:956–961.

Meisinger, J.J. 1984. Evaluating plant-available nitrogen in soil-crop systems. p. 391–416. *In* R.D. Hauck (ed.) Nitrogen in crop production. ASA, CSSA, and SSSA, Madison, WI.

Miller, R.H. 1974. Factors affecting the decomposition of an anaerobically digested sewage sludge in soil. J. Environ. Qual. 3:376–380.

Moorhead, K.K., D.A. Graetz, and K.R. Reddy. 1987. Decomposition of fresh and anaerobically digested plant biomass in soil. J. Environ. Qual. 16:25–28.

Morrill, L.G., and J.E. Dawson. 1967. Patterns observed for the oxidation of ammonium to nitrate by soil organisms. Soil Sci. Soc. Am. Proc. 31:757–760.

Müller, M.M., V. Sundman, O. Soininvaara, and A. Meriläinen. 1988. Effect of chemical composition on the release of nitrogen from agricultural plant materials decomposing in soil under field conditions. Biol. Fertil. Soils 6:78–83.

Myers, R.J.K. 1975. Temperature effects on ammonification and nitrification in a tropical soil. Soil Biol. Biochem. 7:83–86.

Neild, R.E., and M.W. Seeley. 1977. Growing degree days predictions for corn and sorghum development and some applications to crop production in Nebraska. Nebraska Agric. Exp. Stn. Res. Bull. 280. Lincoln.

Olson, R.A., and L.T. Kurtz. 1982. Crop nitrogen requirements, utilization, and fertilization. p. 567–604. *In* F.J. Stevenson (ed.) Nitrogen in agricultural soils. Agron. Monogr. 22. ASA, CSSA, and SSSA, Madison, WI.

Panganiban, E.H. 1925. Temperature as a factor in nitrogen changes in the soil. J. Am. Soc. Agron. 17:1–31.

Parker, D.T., and W.E. Larson. 1962. Nitrification as affected by temperature and moisture content of mulched soils. Soil Sci. Soc. Am. Proc. 26:238–242.

Parr, J.F. 1973. Chemical and biochemical considerations for maximizing the efficiency of fertilizer nitrogen. J. Environ. Qual. 2:75–84.

Paul, E.A., and F.E. Clark. 1989. Soil microbiology and biochemistry. Academic Press, San Diego, CA.

Pierce, F.J., and C.W. Rice. 1988. Crop rotation and its impact on efficiency of water and nitrogen use. p. 21–42. *In* W.L. Hargrove (ed.) Cropping strategies for efficient use of water and nitrogen. ASA Spec. Publ. 51. ASA, CSSA, and SSSA, Madison, WI.

Reamur, R.A.F. 1735. Temperature observations in Paris during the year 1735, and the climatic analogue studies of i'Isle de France, Algeria and some islands of America. (in French). Mem. Acad. Sci. Paris 1735:545.

Ross, D.J., and B.A. Bridger. 1978. Influence of temperature on biochemical processes in some soils from tussock grasslands: 2. Nitrogen mineralization. N.Z. J. Sci. 21:591–597.

Russel, J.C., E.G. Jones, and G.M. Bahrt. 1925. The temperature and moisture factors in nitrate production. Soil Sci. 19:381–398.

Russelle, M.P., W.W. Wilhelm, R.A. Olson, and J.F. Power. 1984. Growth analysis based on degree days. Crop Sci. 24:28–32.

Schimel, D.S., D.C. Coleman, and K.A. Horton. 1985. Soil organic matter dynamics in paired rangeland and cropland toposequences in North Dakota. Geoderma 36:201–214.

Soil Survey Staff. 1975. Soil taxonomy: A basic system of soil classification for making and interpreting soil surveys. USDA-SCS Agric. Handb. 436. U.S. Gov. Print. Office, Washington, DC.

Stanford, G., J.N. Carter, D.T. Westermann, and J.J. Meisinger. 1977. Residual nitrate and mineralizable soil nitrogen in relation to nitrogen uptake by irrigated sugarbeets. Agron. J. 69:303–308.

Stanford, G., and E. Epstein. 1974. Nitrogen mineralization–water relations in soils. Soil Sci. Soc. Am. Proc. 38:103–107.

Stanford, G., M.H. Frere, and D.H. Schwaninger. 1973. Temperature coefficient of soil nitrogen mineralization. Soil Sci. 115:321–323.

Stanford, G., M.H. Frere, and R.A. Vander Pol. 1975. Effect of fluctuating temperatures on soil nitrogen mineralization. Soil Sci. 119:222–226.

Stanford, G., and S.J. Smith. 1972. Nitrogen mineralization potentials of soils. Soil Sci. Soc. Am. Proc. 36:465–472.

Stott, D.E., H.F. Stroo, L.F. Elliott, R.I. Papendick, and P.W. Unger. 1990. Wheat residue loss from fields under no-till management. Soil Sci. Soc. Am. J. 54:92–98.

Tabatabai, M.A., and A.A. Al-Khafaji. 1980. Comparison of nitrogen and sulfur mineralization in soils. Soil Sci. Soc. Am. J. 44:1000–1006.

Tyler, K.B., F.E. Broadbent, and G.N. Hill. 1959. Low-temperature effects on nitrification in four California soils. Soil Sci. 87:123–129.

U.S. Department of Agriculture. 1941. Climate and man. Yearbook of agriculture. U.S. Gov. Print. Office, Washington, DC.

Wolf, J.K., and A.S. Rogowski. 1991. Spatial distribution of soil heat flux and growing degree days. Soil Sci. Soc. Am. J. 55:647–657.

5 Evaluating Potential Nitrogen Mineralization for Predicting Fertilizer Nitrogen Requirements of Long-Term Field Experiments

C. A. Campbell, Y. W. Jame, and O. O. Akinremi
Agriculture and Agri-Food Canada
Swift Current, Saskatchewan, Canada

H. J. Beckie
Agriculture and Agri-Food Canada
Melfort, Saskatchewan, Canada

Nitrogen is one of the most important plant nutrients; it is a major constituent of all plants (Campbell, 1978). Crops obtain N primarily from the mineralization of soil organic matter, but also from added fertilizers, manures, and from legumes (Campbell, 1978). There is an urgent need to develop accurate methods for quantifying the rate at which soils will mineralize N. A solution to this problem would permit more efficient use of N fertilizer, thereby allowing maximization of net returns and reduction of environmental pollution.

Most North American soils were inherently rich in organic matter (Campbell, 1978). Soil degradation and failure to replace adequately the N exported in food, however, has resulted in large N deficits (Fig. 5–1), which has led to a diminution of the N-supplying capacity of these soils, and necessitated an increase in N fertilizer requirements (Campbell et al., 1986). With the increasing use of fertilizers has come an ever-increasing concern by society regarding possible negative influence of N on the wholesomeness of water and air and this has spurred the scientific community to seek methods that producers may adopt so as to manage N in a more responsible and efficient manner.

Generally, practices that result in greater uptake of N by plants and prevent the accumulation of large amounts of mineral N in the soil will reduce the incidence of N pollution of the environment (Bock & Hergert, 1991; Campbell et al., 1994). The objective of this chapter was to demonstrate how the potentially mineralizable N concept (Stanford & Smith, 1972; Stanford et al., 1973; Stanford & Epstein, 1974) may be used in conjunction with deterministic models to estimate net N mineralization in long-term cropping systems. Further, we suggest how it

---

Copyright © 1994 Soil Science Society of America, 677 S. Segoe Rd., Madison, WI 53711, USA.
Soil Testing: Prospects for Improving Nutrient Recommendations, SSSA Special Publication 40.

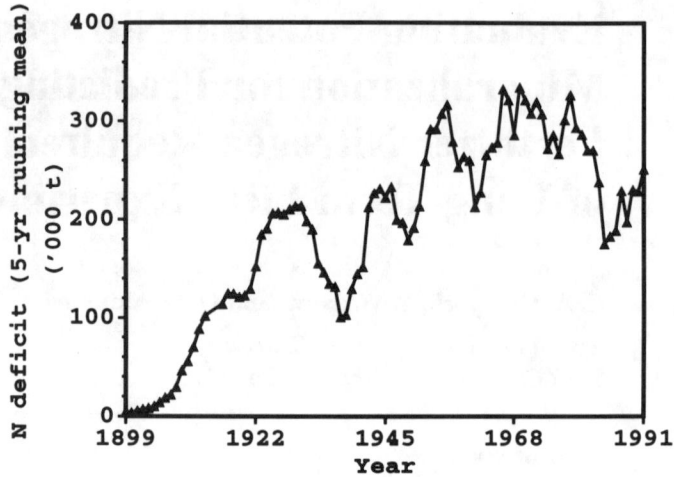

Fig. 5–1. Nitrogen deficit (i.e., grain N minus fertilizer N) for Saskatchewan for period 1899 to 1991 (adapted from Curtin et al., 1994).

may be possible to couple the latter approach with rapid chemical methods of extracting available soil N so as to improve estimation of fertilizer N requirements.

DETERMINING NITROGEN FERTILIZER REQUIREMENTS USING NITROGEN BALANCE

The amount of N fertilizer that must be applied depends on the attainable dry matter yield (Y_{dm}), the N concentration of the crop (N_y), the amount of residual mineral N in the soil at seeding (N_{rm}), and the amount of N that the soil will mineralize from organic matter during the growing season (N_s) (Stanford, 1973). The crop does not recover all of the fertilizer N (N_f) nor soil N, available to it (i.e., efficiency (E) of N_f, N_{rm} and N_s <100%). Therefore, E will influence the quantity of N that must be applied in order to achieve a certain yield. If plant N at harvest is given by $Y_{dm} \times N_y$, then:

$$Y_{dm} N_y = E_{rm} N_{rm} + E_s N_s + E_f N_f \qquad [1]$$

Rearranging:

$$N_f = [Y_{dm} N_y - E_{rm} N_{rm} - E_s N_s] / E_f \qquad [2]$$

A target yield can be set, N_{rm} in the soil at seeding measured, and reasonable estimates of E made, even though the latter will vary depending on weather conditions and management factors (Stanford, 1973; Rennie et al., 1993). The greatest difficulty in solving Eq. [2] is quantifying the N-supplying capacity of the soil during the growing season.

Long before fertilizer use became commonplace, scientists were seeking quick, routine tests for quantifying the N-supplying capacity of soils so as to adjust fertilizer N requirements (Campbell, 1978; Stanford, 1982). This need is even more pressing today because, as producers adopt better management principles, they are likely gradually increasing the N-supplying capacity of their soils (Power et al., 1986; Campbell et al., 1992, 1993b; Mason & Rowland, 1992). For example, in southern Saskatchewan, a rotation of spring wheat (*Triticum aestivum* L.)–lentil (*Lens culinaris* Medikus) grown for 12 yr showed a gradual decrease in requirement for N fertilizer compared with a continuous wheat system that was fertilized based on soil tests (Table 5–1). Similarly, Campbell et al. (1993b) showed evidence that good management practices, including continuous spring wheat grown on zero-tillage and fertilized adequately during 9 yr, had built up the N-supplying capacity of the soil so that in the later years of the study the crop was using soil N more efficiently than fertilizer N. It is unlikely that the empirical NO_3-test that is commonly used on the Canadian prairies and in some regions of the USA to determine fertilizer N requirements will adequately reflect such changes in soil fertility. We presently have no simple soil tests that will allow quantification of such changes in soil fertility; however, later in this chapter we suggest a possible solution.

Since routine tests for N-supplying capacity of soils is being discussed elsewhere in this publication, we will not discuss this further, except to reiterate that most such indices have proven inadequate because they do not measure the potential of the soil to mineralize N during the growing season, nor do they allow us to quantify N_s response to weather conditions.

Table 5–1. Influence of grain legume on trend in fertilizer N requirements for spring wheat in two continuously cropped systems receiving N and P based on soil tests during a 12-yr period† at Swift Current, Saskatchewan (from Campbell et al., 1992).

Year	N fertilizer applied to rotation phase indicated by parenthesis (kg ha^{-1})		
	Cont (W)	W-(Len)‡	Cont (W) minus W-(Len)
1979	39	50	-11
1980	61	61	0
1981	33	28	5
1982	27	27	0
1983	20	21	-1
1984	44	11	33
1985	11	5	6
1986	11	5	6
1987	41	3	38
1988	56	4	52
1989	9	2	7
1990	17	0	17

† Prior to commencement of this experiment this land was cropped continuously to spring wheat fertilized with N and P for 12 yr.
‡ The comparison was made between Cont W and the lentil phase of W-Len because the N requirements of the lentil was dictated by the wheat phase of the rotation.

ESTIMATING NITROGEN MINERALIZED IN SOIL

Although there are several thousand kilograms N per hectare in most agricultural soils of temperate climates, only 1 to 2% of this becomes available to crops each year (Campbell, 1978). The N released during organic matter decomposition is derived from a heterogeneous pool of components varying in stability (Campbell, 1978). These components include fresh crop and animal residues, microbial biomass, microbial metabolites and cell wall constituents adsorbed to colloids, and the very stable humus. Although fresh residues and microbial biomass are important as mineralizable substrates, specific compounds in the mineralizable pool have not been directly identified.

As discussed elsewhere in this book by Cabrera et al. (1994), Stanford and co-workers (Stanford & Smith, 1972; Stanford et al., 1973; Stanford & Epstein, 1974) advanced the concept of potentially mineralizable N, denoted as N_0, and a related mineralization rate constant (k), for use in characterizing soil-available N. Since then, this concept has been used, modified, and discussed in great detail by numerous scientists (Campbell, 1978; Campbell et al., 1981, 1988; Myers et al., 1982; Olness, 1984; Juma et al., 1984; Deans et al., 1986; Bonde & Rosswall, 1987; Paustian & Bonde, 1987; Ellert & Bettany, 1988).

Stanford and Smith (1972) suggested that N_0 and k could be estimated by statistical techniques if the soil was incubated at optimum soil water content and temperature conditions and N_s and time of incubation (t) measured. They assumed that N mineralization under these conditions followed first-order kinetics, i.e.,

$$N_s = N_0 [1 - e^{-kt}] \qquad [3]$$

Under natural conditions N_s must be adjusted for suboptimal environmental conditions in order to predict the actual amount of N mineralized. For North American soils, optimum temperature for N mineralization is assumed to be ~35°C and optimum soil water content is field capacity (−0.03 to −0.01 MPa). The k is temperature-dependent (Stanford et al., 1973; Campbell et al., 1981; 1984). Stanford et al. (1973) suggested Q_{10} (the temperature coefficient) = 2, implying that the rate of N mineralization doubles per 10°C rise in temperature; but Campbell et al. (1984) showed that this value varied with climate. Methods of adjusting N_s for soil water content have been suggested by Stanford and Epstein (1974) and later modified by Myers et al. (1982).

PREDICTING NITROGEN FERTILIZER REQUIREMENTS USING POTENTIAL MINERALIZABLE NITROGEN CONCEPT

In Eq. [2], grain yield goals can be set, straw yields can be easily estimated from well established grain/straw ratios, N_{rm} at seeding can be measured, and we can assume reasonable values for efficiency of use of soil and fertilizer N (Stanford, 1973). To estimate the fertilizer N required (N_f), we must still estimate N_y at harvest, and N_s during the growing season. Stanford (1973) analyzed data taken from the literature and concluded that "despite the wide range of conditions represented, it appears that the percentage of N in total dry matter (aboveground) at maximum attainable yield is unaffected by crop variety, location, climate, or level of attainable yield." He found this value of N for corn (*Zea mays* L.), wheat (*Triticum aestivum* L.), grain sorghum [*Sorghum bicolor* (L.) Moench] and other small grains to average ≈1.12%. Campbell tested this plant N concentration hypothesis using his own data from long-term spring wheat-fertilizer N experiments, however, and obtained values of 1.08% in one instance, but 1.68% in another (data not shown). Nonetheless, because dry matter is by far the main determinant of N content (Clarke et al., 1990), such differences in N concentration might be inconsequential when compared with the uncertainties in other assumptions we often make (e.g., efficiencies), and spatial variability in the field.

Stanford (1973) suggested that the potentially mineralizable N concept could be used to derive N_s. Thus, in Eq. [2], he suggested that N_s be replaced by N_0 $(1 - e^{-kt})$. This would allow us to modify k as a function of temperature and soil water, as done by Campbell et al. (1984, 1988), who used this type of model to predict net N_s under field conditions in a growing season with some success (Fig. 5–2). This type of model has been criticized for various shortcomings (Campbell et al., 1993a), but we believe that most of these are either not quantitatively significant (Campbell et al., 1988), or can be overcome if proper incubation and calculation procedures are employed (Campbell et al., 1993a). For example, all soil samples should be air dried and stored near freezing until analyzed. The incubation temperature is best at 35°C, and the length of incubation should be at least 24 wk (Ellert, 1990). When the model proposed by Campbell et al. (1988) is used for estimating N_s, a factor that can lead to low estimates of N_s is the failure to account for the N accruing from decomposition of the most recent crop residues; but this can be estimated (Campbell et al., 1988), or modeled (e.g., LEACHM and CERES). The model, however, proved least accurate under conditions where a dry soil was wetted (Fig. 5–2, bottom). This type of situation occurs frequently in semiarid climates; thus, this facet requires further research.

For the approach suggested by Stanford (1973) to be used, values of N_0 must be known for each soil. Further, it is desirable that these values do not change markedly from year to year. Few experiments have been conducted in which N_0 has been measured at regular intervals. In one such study carried out at Lethbridge, Alberta, N_0 in the 0 to 0.15 m of soil was determined to be 116, 151, and 128 kg ha^{-1} in 1973, 1978, and 1983, respectively (Dr. Chi Chang, Agriculture Canada Research Station, Lethbridge, 1993, personal communication). In Australia, Dalal and Mayer (1987) found that N_0 decreased rapidly, from 200 to 105 mg kg^{-1} soil during the first 20 yr after breaking and cultivating a Waco soil; however, during the following 50 yr of cultivation N_0 was essentially constant. The advantage of using the model suggested by Stanford is that it allows us to calculate the probable N_s using historical, long-term records of temperature and precipitation (the latter can easily be converted to soil water if soil texture is known). Other characteristics that are required to allow estimation of N_s for any soil are field capacity and wilting point, and if these are not available they can also be estimated from soil texture (Tietje & Tapkenhinrichs, 1993).

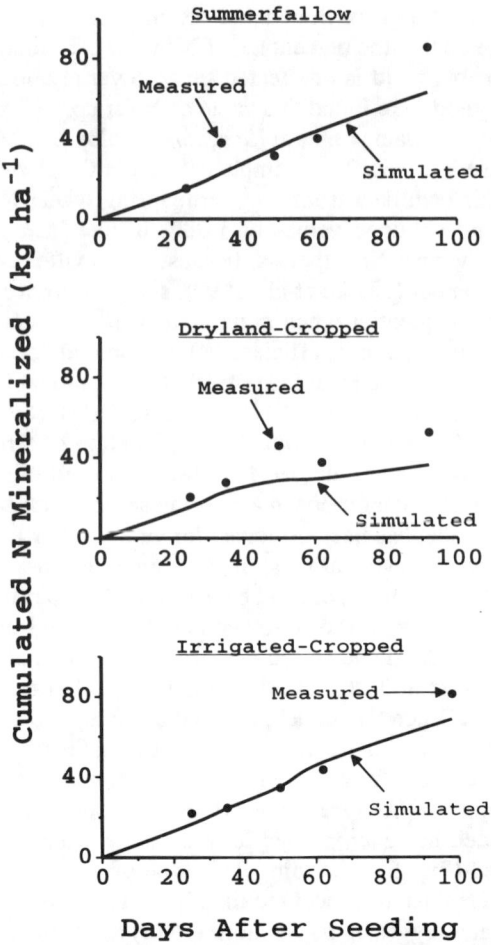

Fig. 5-2. Simulated N mineralization over a growing season compared with values measured in a small lysimeter experiment for three cropping systems at Swift Current, Saskatchewan (adapted from Campbell et al., 1988).

USING POTENTIAL MINERALIZABLE NITROGEN TO ESTIMATE NITROGEN MINERALIZATION IN LONG-TERM ROTATION STUDIES

The concept of potentially mineralizable N has been tested under controlled conditions and in the field (Stanford et al., 1973; Oyanedel & Rodriguez, 1977; Smith et al., 1977; Campbell et al., 1984, 1988) with reasonable success in most cases. But, these have been one-season studies. To be useful in a practical sense, this concept must perform credibly for systems that may be degrading or aggrading as a consequence of management.

To test the usefulness of this concept under natural field conditions over several years requires the use of dynamic models that will allow accounting for the main components of the N cycle. Numerous models of this type exist (e.g., EPIC, NTRM, NLEAP, CERES, and LEACHM). At Swift Current, we have tested the ability of LEACHM and CERES to simulate N mineralization in two rotations of a long-term crop rotation study (Campbell et al., 1983b, 1992). We used the NO_3 leaching version of LEACHM (i.e., LEACHMN) to model N dynamics in the well-fertilized continuous spring wheat (Cont W) and fallow-spring wheat (F-W) rotations during the period 1967 to 1991 (Akinremi et al., 1993). In the case of CERES, we only simulated a wet and a dry year. These two models simulate most processes of the N cycle in a similar manner, but LEACHM does a more rigorous job of modeling solute movement in soil, while CERES is more effective in modeling plant growth and N uptake (Hutson & Wagenet, 1991; Jame et al., 1993). In these two models, we found the N subroutines (Hutson & Wagenet, 1991) to be inadequate; consequently, we modified these N submodels by splitting the humus component into a slow N release fraction (humus), and an active component (N_0 in Fig. 5–3) with characteristics previously determined for this soil (Campbell et al., 1984, 1988). The results obtained after our modification of these two models were much more plausible than when the unmodified models were used.

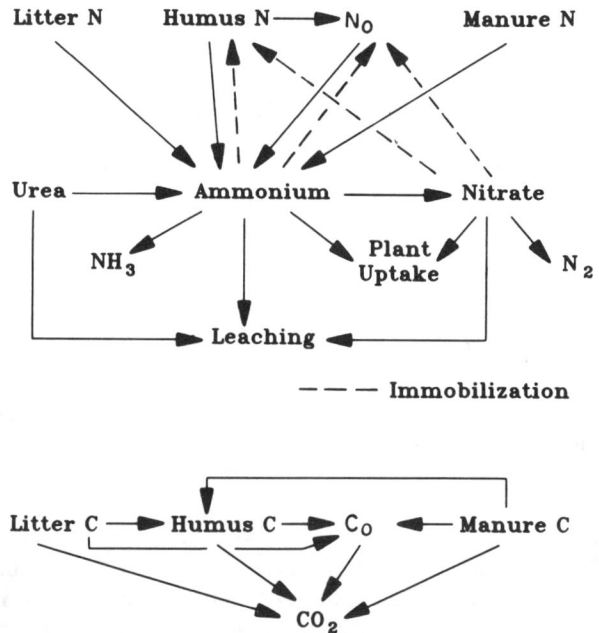

Fig. 5–3. Modified N submodel of LEACHMN (Hutson & Wagenet, 1991) showing potentially mineralizable nitrogen (N_0) and potentially mineralizable carbon (C_0) (adapted from Akinremi et al., 1993).

SIMULATION WITH LEACHMN MODEL

Before using LEACHMN to analyze the rotation data from Swift Current we modified several subroutines (Akinremi et al., 1993) and then calibrated it using data gathered in 1976 in an experiment conducted to measure Cl movement in lysimeters in the field (Campbell, 1977, unpublished data). The model was validated using data from another lysimeter experiment (Campbell et al., 1977). The results obtained for water, Cl and NO_3 distribution in the soil profile throughout the growing season, under summer fallow and cropped systems, indicated a realistic representation of the system had been achieved using the modified model (Akinremi et al., 1993).

We used this modified, calibrated model to simulate water and N in the rotation study. Total soil N (0.185%) assumed to be the starting value (Biederbeck et al., 1984) was used to initialize the model for the year 1967. This was split into two fractions, an active fraction (N_o) and a stable fraction (humus-N) based on the ratio of N_o/total N measured on the same soil in 1982 (Campbell et al., 1984). In that study, the ratio of N_0 to total N was found to be 0.08, hence the initial value of total N in 1967 of 0.185 (or 1850 mg kg^{-1}) was split into 148 mg kg^{-1} as N_0 and 1702 mg kg^{-1} as humus in the 0- to 0.15-m layer. A similar calculation was made for the 0.15- to 0.30-m depth. For C, this was initialized using the value of organic C measured in 1967 and assuming the C/N ratio of the potentially mineralizable fraction was 10. The value of the potentially mineralizable carbon (C_0) was subtracted from the total C to arrive at the humus C at the beginning of the experiment. It was assumed that very little N mineralization occurred below the 0.30–m depth (Campbell & Biederbeck, 1982).

The k of the potentially mineralizable N for the 0- to 0.15-m layer, taken from Campbell et al. (1984) for laboratory incubation measurements carried out on a similar soil at 35°C, was 0.013 d^{-1}. The k value for the 0.15- to 0.30-m layer was arbitrarily assumed to be one-tenth of that for the surface 0.15 m. It was assumed that the k did not vary between growing seasons nor during a growing season.

The amount of straw-N available at the start of the experiment was unknown, but it was estimated from straw yield of an adjacent field in 1966 (Campbell et al., 1969). During the experiment the actual straw dry wt and straw N were used in the model calculations. The straw plus roots was assumed plowed into the soil and made available for decomposition a day after harvest. Root dry matter was estimated from the grain and straw dry matter (Campbell et al., 1977). Straw and root C concentration were assumed to be 45% and root N concentration 1% (Campbell et al., 1991a). The measured N concentrations of the straw were used.

When the modified model was used to simulate water and N disposition in the soil in the two designated crop rotations at Swift Current, the model performed very well for water (data not shown) and reasonably well for NO_3 (Fig. 5–4). As is readily seen in this figure, even in this apparently uniform site, spatial variability of NO_3–N in the field can be quite large, especially in the 0.6- to 1.2-m depth. Therefore, striving for extreme precision in these types of analyses does not seem warranted.

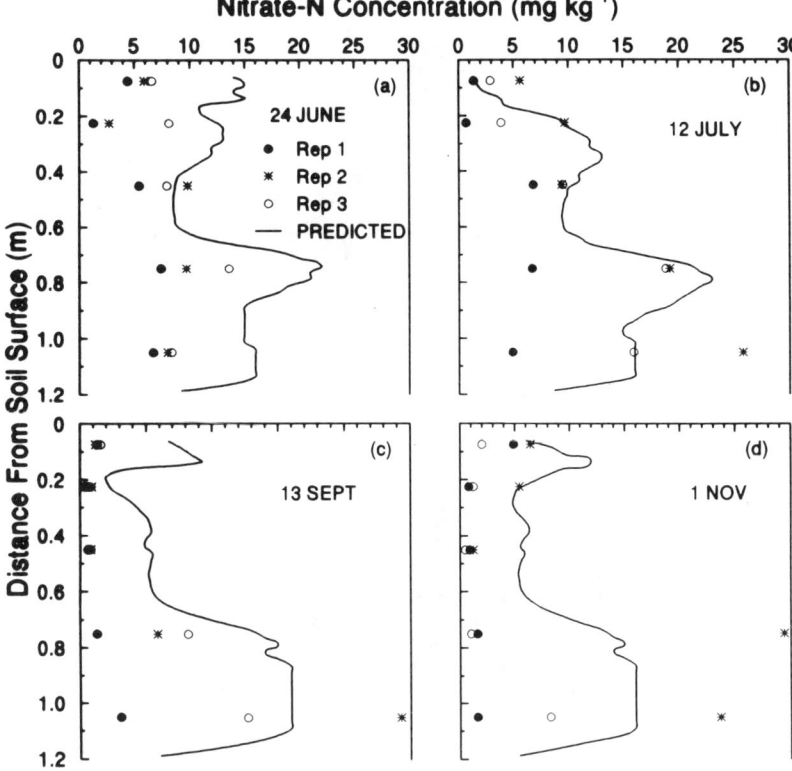

Fig. 5–4. Simulated soil NO_3 distribution under fallow phase of F-W rotation in 1982 (wet year). Rainfall from 6 May to 24 June = 110 mm, 24 June to 12 July = 84 mm, 12 July to 13 September = 94 mm, and 13 September to 1 November = 67 mm (adapted from Akinremi et al., 1993).

The incorporation of the concept of potentially mineralizable N into LEACHMN allowed us to estimate changes in N_0 during the growing season. Few studies have been conducted that allow this assessment. Bonde and Rosswall (1987) reported that N_s (and presumably N_0) decreased during the growing season, but increased again after harvest, as a result of crop residue input. This was similar to the results we obtained for the Cont W system in the wet 1982 (Fig. 5–5, top), which favored N-mineralization. Consequently, the model predicted that N_0 declined gradually during the growing season until harvest (13 September), then N_0 increased in response to the fresh residues. But, as shown by the results for the same rotation in 1978 (Fig. 5–5, bottom), the pattern of change in N_0 during a growing season depends upon the amount of crop residue obtained in the previous year. Since N_0 is reduced by mineralization and increased through immobilization, the process that predominates during the growing season will dictate the trend of N_0. In years where crop residues are low, the build-up of N_0 may not compensate for its decline. In such situations N_0 will decline throughout the growing season until harvest when input of crop residues occur.

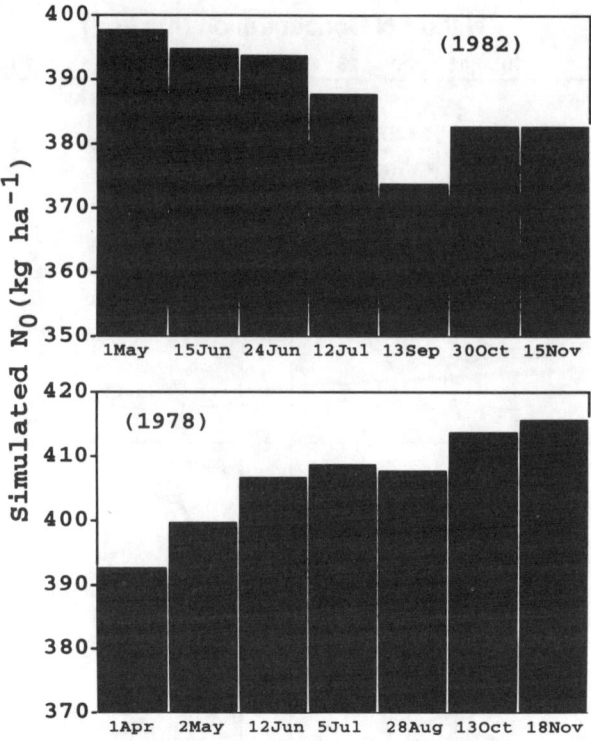

Fig. 5–5. Changes in simulated potential mineralizable nitrogen (N_0) in the 0- to 0.3-m depth, under Cont W during the growing season of two dissimilar years at Swift Current, Saskatchewan (adapted from Akinremi et al., 1993).

The simulations predicted that, during the 24-yr period, N_0 will decline by 40% under F-W, but by only 10% under Cont W (Fig. 5–6). The rapid decline in N_0 under both rotations in the first 4 yr of the experiment was due to low yields during the first 3 yr. While N_0 recovered in Cont W after the fourth year, N_0 in the F-W rotation continued to decline. The higher annual production of crop residues under Cont W and the rapid decomposition of residues during the fallow phase of the 2-yr rotation account for this difference. These results are in accord with reports in the literature (Campbell et al., 1991a,b). Humus-N was predicted to decline by ~2% under Cont W and by ~4% under F-W during the 24-yr period. These amounts of loss are small compared with the total amounts of humus-N. [Note that the rate constant (k_h) for the mineralization of humus-N was not measured; it was obtained through calibration of the model (k_h = 0.00083 d^{-1})]. Although the humus fraction declined very little during the period simulated, the model suggests that the contribution of this fraction to the mineralized N was very important. In several years, mineral N produced from humus-N constituted ~50% of net mineralized N because of the magnitude of the humus fraction. Paul and Juma (1981) reported from studies in which ^{15}N techniques were used to study mineralization–immobilization of N, that the relative contribution of

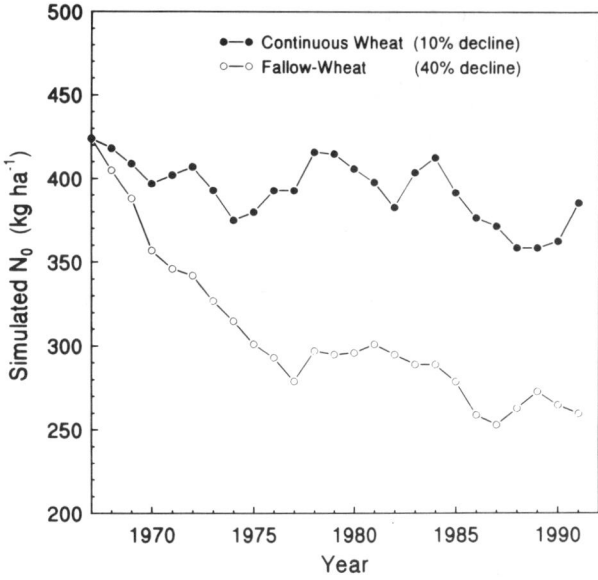

Fig. 5–6. Trends in simulated potentially mineralizable N (N_0) during 24 yr under F-W and Cont W rotations at Swift Current, Saskatchewan (values are for the 0- to 0.3-m depth; adapted from Akinremi et al., 1993).

various N fractions to the mineralized-N pool were: biomass 24%, metabolites 4%, active-N 32%, and stabilized-N, 40%. The first three pools would be approximately equivalent to N_0 in our model. The relative contribution of humus-N to the mineral-N fraction, as predicted by this modified LEACHMN model, was therefore in reasonable agreement with that reported by Paul and Juma.

The simulated amount of N_s for the period between spring thaw and freeze up each year was estimated for Cont W, the fallow phase of F-W [(F)-W], and the wheat phase of F-W, [F-(W)] (Table 5–2). Each phase of F-W was present each year, thus there is a value for each phase each year. The results for F-(W) and Cont W seem reasonable compared with estimates of N_s made by balance sheet methods (Campbell & Paul, 1978; Campbell et al., 1983a, 1992). Values for (F)-W were lower than for F-(W), which does not seem reasonable. We expected the (F)-W, with a longer period of moist soil conditions, to accumulate the most mineral N in this period. The F-(W) system should accumulate more N_s than Cont W in years when the previous fall and winter were dry such that stored water in surface soil of fallow exceeded that in surface soil of Cont W; otherwise, both of these systems should behave similarly. The simulation results support the latter hypothesis. Generally, the magnitude of the simulated N_s is in the range estimated for cereal systems in this area (Nyborg et al., 1976; Campbell & Paul, 1978; Campbell et al., 1974, 1988, 1992). The reason for the apparent inadequacy in simulating the (F)-W system could be related to the fact that all the residues from the previous years' crop are assumed to be available for decomposition immediately after harvest. This could result in overestimation of N immobilization.

Table 5–2. Net N mineralized (N_s) from spring thaw to wheat harvest for fallow-wheat (both phases) and continuous wheat as estimated by simulation with the LEACHMN model (from Akinremi et al., 1993).

Year	Net N_s in (F)-W†		Net N_s in F-(W)†		Cont W
	kg ha⁻¹				
1967	—	47	27	—	27
1968	26	—	—	40	18
1969	—	41	42	—	34
1970	48	—	—	55	40
1971	—	28	37	—	22
1972	33	—	—	29	23
1973	—	22	36	—	36
1974	34	—	—	41	47
1975	—	23	40	—	27
1976	26	—	—	43	23
1977	—	25	46	—	25
1978	18	—	—	36	5
1979	—	27	30	—	24
1980	26	—	—	37	30
1981	—	24	33	—	32
1982	43	—	—	48	49
1983	—	15	36	—	9
1984	21	—	—	21	5
1985	—	42	28	—	37
1986	41	—	—	37	43
1987	—	33	33	—	27
1988	20	—	—	27	30
1989	—	37	29	—	37
1990	40	—	—	32	24
1991	—	22	45	—	15
Mean	31	30	36	37	28

†Each phase of the fallow-wheat rotation was present each year. The values pertain to the phase in parenthesis. (F)-W means N mineralized during fallow phase; F-(W) means N mineralized during wheat phase.

SIMULATIONS WITH CERES MODEL

The CERES model was calibrated and modified especially for use on the Canadian Prairies (Jame et al., 1993). We used this modified version of CERES to estimate N_s during a wet year (1982) and a dry year (1973) at Swift Current, for F-W and Cont W.

The crop was seeded on 15 May in 1973 and on 5 June in 1982; however, we simulated periods from spring soil sampling (3 May 1973 and 6 May 1982) to the date of crop physiological maturity (9 Aug. 1973 and 3 Sept. 1982). The soil water contents in the profile (0–1.2 m), measured at the soil sampling date, were used as the initial values in the simulation. The initial soil water contents were much higher in 1973 (250–270 mm) than in 1982 (131–212 mm). Total precipitation received during the simulation periods were 62 mm for 1973 and 310 mm for 1982.

In CERES, the mineralization and immobilization subroutine simulates the decay of two types of organic matter: fresh organic matter (FOM, which includes crop residues or green manure), and a more stable organic or humic pool (HUM). The FOM is further subdivided into three pools: carbohydrate (20%), cellulose (70%), and lignin (10%). The first-order rate of decay constants used in the model were 0.2, 0.05, and 0.0095 d^{-1} for carbohydrate, cellulose, and lignin, respectively. Input data for this model includes, the amount of straw added, its C/N ratio, and its depth of incorporation into soil and an estimate of the amount of root residue from the previous crop. These data are used to initialize FOM and the N contained within the FON for each soil layer.

In the simulation, we assumed that 70% of the straw produced in the previous crop year was present at the start of the simulation run for (F)-W and cont (W) (cropped the previous year) and 15% for F-(W) (cropped 2 yr previously). We also assumed that the straw was uniformly incorporated within the 0.1-m layer during seedbed preparation. Root dry matter was estimated as 15% of the straw weight. The C concentration of straw and root was assumed to be 45%, and N concentration of the straw assumed to be 0.6%.

The mineralization subroutine in the CERES model requires that the soil organic C in each soil layer be specified. This is used to calculate the initial humic pool for each layer and, together with a simplifying assumption of bulk soil C/N ratio of 10, to estimate the N associated with the humic pool (NHUM). The CERES model also allows for the transfer of 20% of the gross amount of N mineralized each day from FON pool into the humic pool. As organic matter decomposes, some N is required by the decay process and this amount is incorporated into microbial biomass that is included in the FON pool. The balance between the N immobilized and the N mineralized from FON and NHUM determines whether net mineralization or immobilization occurs.

In the CERES model, the decay rate constant for calculating the N released from the humus does not vary with depth. Thus, the amount of N mineralized is calculated from the whole profile, but the contribution of N mineralized from the deeper soil layers is very small because of the associated low values of soil organic C content.

Without any modification of the CERES model, the largest amount of net N mineralized during the growing season from the six cropping systems simulated was only 21 kg N ha^{-1} [(F)-W in 1982] and the smallest amount was 5 kg N ha^{-1} for Cont (W) in 1973 (Fig. 5–7 and 5–8). These values are substantially lower than those generally reported for this soil (Campbell et al., 1974, 1988, 1992; Campbell & Paul, 1978). These low estimates of N mineralization by the unmodified CERES model were mainly associated with the very small k value from humus (k_h = 0.000083 d^{-1}). This value was adopted from the PAPRAN model (Seligman & van Keulen, 1981). In the SOIL-SOILN model, Bergstrom and Jarvis (1991) used a higher rate constant for k_h (0.0003 d^{-1}).

We modified the N submodel of CERES by splitting the humus component into a slow N release fraction (humus) and a more active component (N_0) (Fig. 5–3). The N that is immobilized was incorporated into the N_0 pool instead of the FON pool. Initially, the ratio of N_0 to total N in the humic pool was set to be 0.08 for all soil layers. The rate constant of the potentially mineralizable N for the surface layer (0–0.1 m) was 0.013 d^{-1} as discussed previously, and k_h was 0.000083 d^{-1} as used in the original CERES model. The k for the 0.1- to 0.2-m layer, however, was assumed to be 0.3 of that in the 0.0- to 0.1-m layer and for the 0.2- to 0.4-m layer, it was assumed to be 0.1 of the rate constant of the 0.0- to 0.1-m layer. We assumed that no N mineralization occurred below 0.40 m.

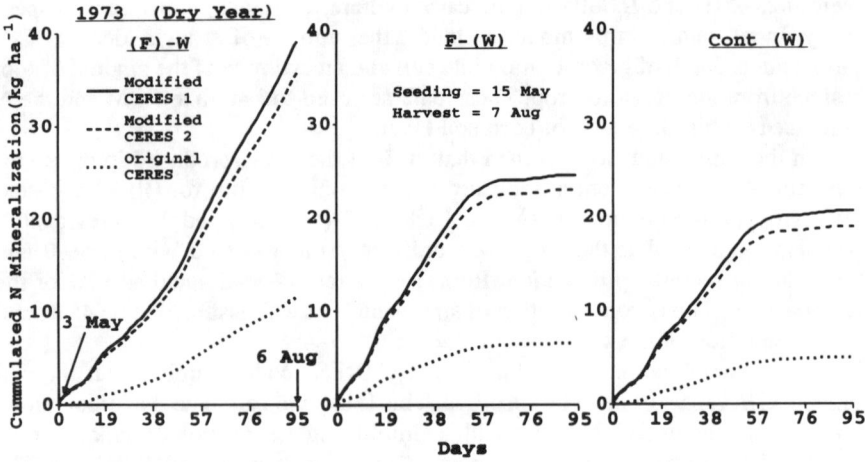

Fig. 5–7. Simulated N mineralized in fallow-spring wheat (F-W) and continuous wheat (Cont W), during the growing season of a dry year (1973), at Swift Current, Saskatchewan. The initial water content in the 0- to 1.2-m soil depth was 248 mm in (F)-W, 262 mm in F-(W), and 270 mm in Cont (W). (Modified CERES 1 refers to case where N_0 and humus pools were used, while modified CERES 2 uses only N_0). Precipitation received between 4 May and 9 August was 62 mm.

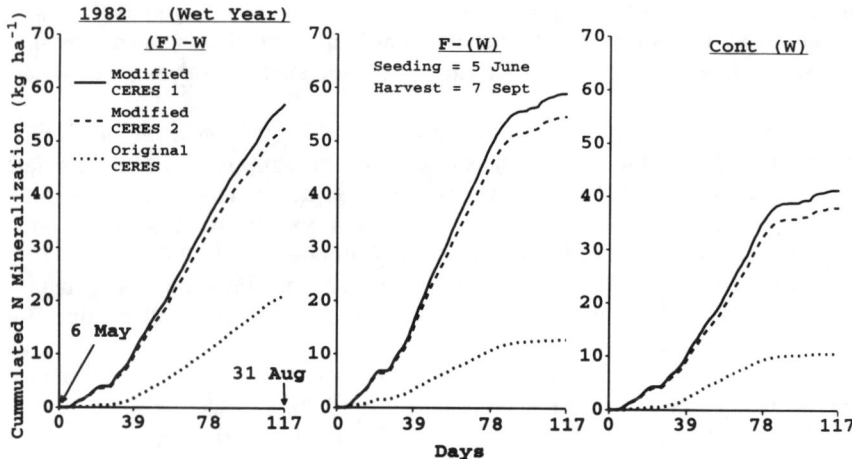

Fig. 5–8. Simulated N mineralized in fallow-spring wheat (F-W) and continuous wheat (Cont W), during the growing season of a wet year (1982), at Swift Current, Saskatchewan. The initial water content in the 0- to 1.2-m soil depth was 131 mm in (F)-W, 212 mm in F-(W), and 163 mm in Cont (W). Precipitation received between 7 May and 2 September was 309 mm. (Modified CERES 1 refers to case where N_0 and humus pools were used, while modified CERES 2 uses only N_0).

When this modified CERES model was used to predict N_s during the growing season, it predicted 39, 25, and 21 kg N ha^{-1} produced by (F)-W, F-(W), and Cont(W), respectively, in the dry 1973 (CERES 1, Fig. 5–7); LEACHMN predicted corresponding values of 22, 36, and 36 kg N ha^{-1} (Table 5–2). In the wet year (1982), the corresponding values for the modified CERES were 58, 59, and 42 kg N ha^{-1} (CERES 1, Fig. 5–8), while LEACHMN predicted 43, 48, and 49 kg N ha^{-1}, respectively. The results obtained with CERES appear to be more plausible, both in terms of the relative size of N_s for (F)–W compared with the other two systems, and in terms of N_s in the wet year compared with the dry year.

We reexamined our premise for revising the LEACHM and CERES N submodels (Fig. 5–3) and reasoned that because the measured N_0 was derived from all available pools of N (except litter, which is usually removed and discarded during incubation), then we should only have replaced the humus with N_0. That is, in Fig. 5–3 we were in essence double-counting the humus fraction. We therefore conducted another simulation with CERES model modified to replace humus with N_0. This simulation resulted in N_s during the growing season being ≈5 kg N ha^{-1} less than in CERES 1 (CERES 2, Fig. 5–7 and 5–8). Thus, even if the revised model (Fig. 5–3) was invalid, the error was not great because k_h in the CERES model is so small (0.000083 d^{-1}).

An estimate of the amount of N mineralized in 1973 from crop residues showed 8.6 kg ha^{-1} for (F)-W, 5.3 for F-(W), and 4.9 for Cont(W). Similar estimates for 1982 were 18.2 kg ha^{-1} for (F)-W, 3.7 for F-(W), and 9.0 for Cont(W).

The validity of the N_s values obtained in these simulations is not easily verified because most balance sheet estimations are crude, involving several gross assumptions. The values obtained by using LEACHMN and CERES, however, were within the range of N_s estimated by balance sheet methods for this soil in several studies. Using these dynamic models, as compared with the simple model used by Campbell et al. (1988), provides an advantage in that they allow us to account for N mineralized from recent crop residues and also to account for N lost by leaching, gaseous means and N immobilized. As shown, these models require further fine-tuning, but the results obtained by introducing the N_0 concept into the N submodels make these models more credible.

RELATING POTENTIALLY MINERLIZABLE NITROGEN TO QUICK ROUTINE METHODS OF MEASURING AVAILABLE NITROGEN FOR SOIL TESTING PURPOSES

In the foregoing discussion we demonstrated that the concept of potentially mineralizable N can be used effectively to quantify the amount of N that a soil will mineralize during a growing season under cropped and fallow systems. Of the factors required to make this estimate, the most difficult to obtain is N_0 because it will differ for each soil (Stanford & Smith, 1972; Campbell et al., 1984) and because its determination requires long-term incubation. The latter would make this technique impractical for use in soil testing laboratories. Estimating the k value is less difficult because we can assume $Q_{10} = 2$ to be a reasonable approximation for most soils.

If this approach is to be used by soil testing laboratories, a quick, effective routine extraction method of estimating N_0 is required. Two promising methods proposed by Gianello and Bremner (1986a,b), involves (i) determination of the NH_4–N produced when the soil is digested with 2 M KCl at 100°C for 4 hr, and (ii) measuring NH_4–N produced by steam distillation of soil with pH 11.2 phosphate-borate buffer solution, for 8 min. Both methods were tested using 33 Brazilian soils and 30 diverse Iowa soils with very encouraging results. For example, these workers obtained excellent agreement between the KCl-extracted N and the amount of N mineralized during the first 14 d of aerobic incubation at 35°C (Stanford & Smith, 1972) for 33 Brazilian soils (Fig. 5–9). Although promising, this relationship is unfortunately not made relative to N_0, which is what is required. If we could relate N_0 to hot KCl-N or phosphate-borate-N, then it should be possible to determine N_0 for major soils of a region, develop the relationship between the chemically extractable N and N_0 in these soils, and store this information in the computer files of soil test laboratories in each region. It should then be possible to determine the chemically extracted N on a producer's soil, once every 3 to 5 yr, and estimate N_0 from the known chemically extractable-N vs. N_0 relationship. All the required characteristics for each producer's fields could be determined once, and stored in computer files for use in estimating fertilizer N requirements each year based on the soil test mineral N, the yield goal, and the degree of risk the producer is willing to take (i.e., in terms of precipitation and temperature). We could then use this information together with an appropriate deterministic model and historical weather data for an area to estimate the probable N_s.

We analyzed data from two previous studies (Stanford & Smith, 1972; Campbell et al., 1984) to determine how close a relationship we might expect to

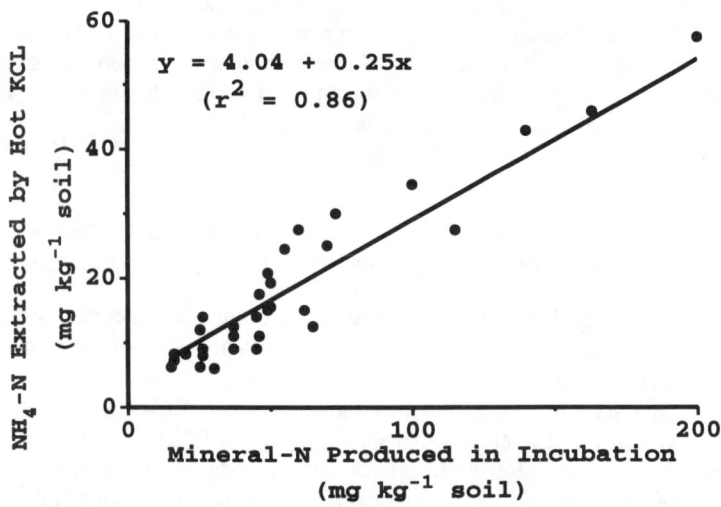

Fig. 5–9. Relationship between NH_4–N value obtained by hot, 2 M KCl extraction, and (NH_4 + NO_3 + NO_2)–N produced by aerobic incubation of soil at 35°C for 14 d (33 soils) (adapted from Gianello & Bremner, 1986).

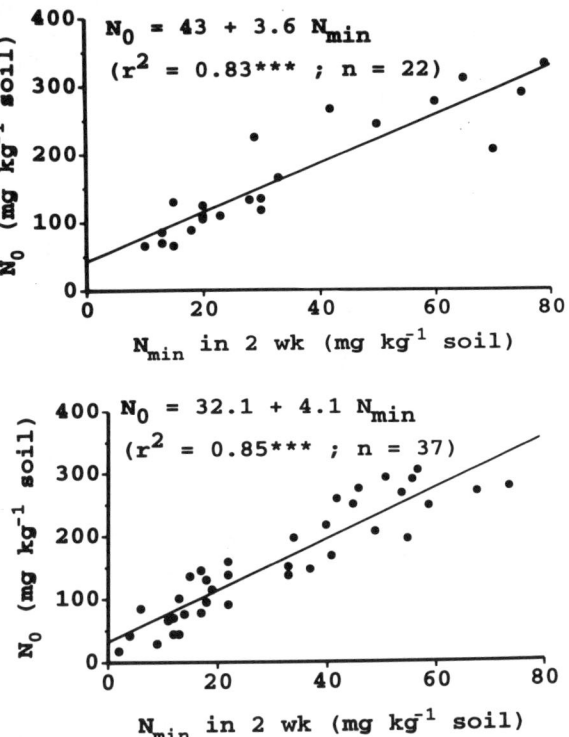

Fig. 5–10. Relationship between N_0 and N mineralized during first 2 wk of aerobic incubation at 35°C. (top) Adapted from Campbell et al. (1984); (bottom) adapted from Stanford and Smith (1972).

find between N_0 and the N_s produced in the 0 to 2-wk period of aerobic incubation. The results (Fig. 5–10) were very encouraging, with ≈85% of the variability in N_0 being accounted for by N mineralized in 0 to 2 wk. This suggests that we are likely to obtain a close relationship between N_0 and the hot KCl or phosphate-borate extractable N; however, this remains to be determined experimentally.

CONCLUSIONS

The need to develop accurate methods for quantifying the rate at which soils will mineralize N cannot be overemphasized. Producers require a solution to this problem so that they can make more efficient use of N fertilizer and thus maximize net returns and reduce environmental pollution. Society demands the latter! Scientists have similar goals to producers and society, but they have the added obligation to enhance their understanding of the mechanisms that control N behavior in the soil–plant–air system.

This chapter has demonstrated that scientists are making significant progress towards achievement of these goals. More specifically, it demonstrates that the potentially mineralizable N concept of Stanford and colleagues can be used, together with deterministic models such as LEACHMN and CERES, to estimate N mineralization in long-term cropping systems. The chapter further shows that it might be possible to couple the potentially mineralizable N concept with a rapid hot KCl or phosphate-borate extractable N procedure of Gianello and Bremner to provide an improved soil test for N. If this latter concept is successfully developed, it offers the advantage over the current soil tests in that it would allow quantification of the capacity of soils to mineralize N while allowing modification in estimates of N mineralization to be made based on probable temperature and soil water for an area.

Much research remains to be done however, (i) to improve the accuracy of determining N_0 and k, (ii) to improve the N submodels of deterministic models such as CERES and LEACHMN, and (iii) to establish whether the Gianello-Bremner chemical procedures are effective on a universal basis.

REFERENCES

Akinremi, O.O., C.A. Campbell, Y.W. Jame, R.P. Zentner, and C. Chang. 1993. Simulating nitrogen dynamics and nitrate leaching using LEACHM model. Publ. 379M0083. Research Branch, Agriculture Canada, Res. Stn., Swift Current, SK.

Bergstrom, L., and N.J. Jarvis. 1991. Prediction of nitrate leaching losses from arable land under different fertilization intensities using the SOIL-SOILN models. Soil Use Manage. 7:79–85.

Biederbeck, V.O., C.A. Campbell, and R.P. Zentner. 1984. Effect of crop rotation and fertilization on some biological properties of a loam in southwestern Saskatchewan. Can. J. Soil Sci. 64:355–367.

Bock, B.R., and G.W. Hergert. 1991. Fertilizer nitrogen management. p. 139–164. *In* F. Follet et al. (ed.) Managing nitrogen for groundwater quality and farm profitability. SSSA, Madison, WI.

Bonde, T.A., and T. Rosswall. 1987. Seasonal variation of potentially mineralizable nitrogen in four cropping systems. Soil Sci. Soc. Am. J. 51:1508–1514.

Cabrera, M.L., M.F. Vigil, and D.E. Kissel. 1994. Potential nitrogen mineralization: Laboratory and field evaluation. p. 15–30. *In* J.L. Havlin and J.S. Jacobsen (ed.) Soil testing: Prospects for improving nutrient recommendations. SSSA Spec. Publ. 40. SSSA, Madison, WI.

Campbell, C.A. 1978. Soil organic carbon, nitrogen and fertility. Dev. Soil Sci. 8:173–272.

Campbell, C.A., and V.O. Biederbeck. 1982. Changes in mineral N and numbers of bacteria and actinomycetes during two years under wheat fallow in southwestern Saskatchewan. Can. J. Soil Sci. 62:125–137.

Campbell, C.A., V.O. Biederbeck, R.P. Zentner, and G.P. Lafond. 1991a. Effect of crop rotations and cultural practices on soil organic matter, microbial biomass and respiration in a thin Black Chernozem. Can. J. Soil Sci. 71:363–376.

Campbell, C.A., D.R. Cameron, W. Nicholaichuk, and H.R. Davidson. 1977. Effect of fertilizer N and soil moisture on growth, N content and moisture use by spring wheat. Can. J. Soil Sci. 57:289–310.

Campbell, C.A., B.H. Ellert, and Y.W. Jame. 1993a. Nitrogen mineralization in soils. p. 341–349. *In* M.R. Carter (ed.) Soil sampling and analytical methods. Lewis Pub., Boca Raton, FL.

Campbell, C.A., Y.W. Jame, and R. DeJong. 1988. Predicting net nitrogen mineralization over a growing season: Model verification. Can. J. Soil Sci. 68:537–552.

Campbell, C.A., Y.W. Jame, and G.E. Winkleman. 1984. Mineralization rate constants and their use for estimating nitrogen mineralization in some Canadian prairie soils. Can. J. Soil Sci. 64:333–343.

Campbell, C.A., G.P. Lafond, A.J. Leyshon, R.P. Zentner, and H.H. Janzen. 1991b. Effect of cropping practices on the initial potential rate of N mineralization in a thin Black Chernozem. Can. J. Soil Sci. 71:43–53.

Campbell, C.A., G.P. Lafond, and R.P. Zentner. 1994. Nitrate leaching in a Udic Haploboroll as influenced by fertilizer and legumes. J. Environ. Qual. 23:195–201.

Campbell, C.A., R.J.K. Myers, and K.L. Weier. 1981. Potentially mineralizable nitrogen, decomposition rates and their relationship to temperature for five Queensland soils. Aust. J. Soil Res. 19:323–332.

Campbell, C.A., and E.A. Paul. 1978. Effect of fertilizer N and soil moisture on mineralization, N recovery, A-values under spring wheat grown in small lysimeters. Can. J. Soil Sci. 58:39–51.

Campbell, C.A., W.L. Pelton, and K.F. Nielsen. 1969. Influence of solar radiation and soil moisture on growth and yield of Chinook wheat. Can. J. Plant Sci. 49:685–699.

Campbell, C.A., D.W.L. Read, V.O. Biederbeck, and G.E. Winkleman. 1983a. First 12 years of a long-term crop rotation study in southwestern Saskatchewan—nitrate–N distribution in soil and N uptake by the plant. Can. J. Soil Sci. 63:563–578.

Campbell, C.A., D.W.L. Read, R.P. Zentner, A.J. Leyshon, and W.S. Ferguson. 1983b. First 12 years of a long-term crop rotation study in southwestern Saskatchewan—yields and quality of grain. Can. J. Plant Sci. 63:91–108.

Campbell, C.A., D.W. Stewart, W. Nicholaichuk, and V.O. Biederbeck. 1974. Effects of growing season soil temperature, moisture, and NH_4–N on soil nitrogen. Can. J. Soil Sci. 54:403–412.

Campbell, C.A., R.P. Zentner, J.F. Dormaar, and R.P. Voroney. 1986. Land quality, trends and wheat production in western Canada. p. 318–353. In A.E. Slinkard and D.B. Fowler (ed.) Wheat production in Canada — A review. Proc. of the Canadian Wheat Production Symposium, Saskatoon, SK. 3–5 Mar. 1986. Div. Ext. and Community Relations, Univ. of Saskatchewan, Saskatoon.

Campbell, C.A., R.P. Zentner, F. Selles, V.O. Biederbeck, and A.J. Leyshon. 1992. Comparative effects of grain lentil-wheat and monoculture wheat on crop production, N economy and N fertility in a Brown Chernozem. Can. J. Plant Sci. 72:1091–1107.

Campbell, C.A., R.P. Zentner, F. Selles, B.G. McConkey, and F.B. Dyck. 1993b. Nitrogen management for spring wheat grown annually on zero-tillage: Yields and N use efficiency. Agron. J. 85:107–114.

Clarke, J.M., C.A. Campbell, H.W. Cutforth, R.M. DePauw, and G.E. Winkleman. 1990. Nitrogen and phosphorus uptake, translocation and utilization efficiency of wheat in relation to environment and cultivar yield and protein levels. Can. J. Plant Sci. 70:965–977.

Curtin, D., F. Selles, C.A. Campbell, and V.O. Biederbeck. 1994. Canadian Prairie agriculture as a source and sink of the greenhouse gases, carbon dioxide and nitrous oxide. Publ. 379M0082. Research Branch, Agriculture Canada, Research Station, Swift Current, SK.

Dalal, R.C., and R.J. Mayer. 1987. Long-term trends in fertility of soils under continuous cultivation and cereal cropping in southern Queensland: VII. Dynamics of nitrogen mineralization potentials and microbial biomass. Aust. J. Soil Res. 25:461–472.

Deans, J.R., A.E. Molina, and C.E. Clapp. 1986. Models for predicting potentially mineralizable nitrogen and decomposition rate constants. Soil Sci. Soc. Am. J. 50:323–326.

Ellert, B.J. 1990. Kinetics of nitrogen and sulfur cycling in Gray Luvisol soils. Ph.D. thesis. Univ. of Saskatchewan, Saskatoon, SK.

Ellert, B.H., and J.R. Bettany. 1988. Comparison of kinetic models for describing net sulfur and nitrogen mineralization. Soil Sci. Soc. Am. J. 52:1692–1702.

Gianello, C., and J.M. Bremner. 1986a. A simple chemical method of assessing potentially available organic nitrogen in soil. Commun. Soil Sci. Plant Anal. 17:195–214.

Gianello, C., and J.M. Bremner. 1986b. Comparison of chemical methods of assessing potentially mineralizable organic nitrogen in soil. Commun. Soil Sci. Plant Anal. 17:215–236.

Hutson, J.L., and R.J. Wagenet. 1991. Simulating nitrogen dynamics in soils using a deterministic model. Soil Use Manage. 7:74–78.

Jame, Y.W., J. Liick, and W.A. Thick. 1993. A user's guide to the modified CERES-wheat model for use on the Canadian Prairies. Research Branch, Agriculture Canada. Swift Current, SK.

Juma, N.G., E.A. Paul, and B. Mary. 1984. Kinetic analysis of net nitrogen mineralization in soil. Soil Sci. Soc. Am. J. 48:753–757.

Mason M.G., and I.C. Rowland. 1992. Effect of amount and quality of previous crop residues on the nitrogen fertilizer response of a wheat crop. Aust. J. Exp. Agric. 32:363–370.

Myers, R.J.K., C.A. Campbell, and K.L. Weier. 1982. Quantitative relationship between net nitrogen mineralization and moisture content of soils. Can. J. Soil Sci. 62:111–124.

Nyborg, M., J.A. Neufeld, and R.A. Bertrand. 1976. Measuring crop available nitrogen. p. 102–127. In Proc. Western Canada Nitrogen Symp., Alberta Soil Sci. Workshop, Calgary. 20–21 Jan. 1976. Alberta Agriculture, Edmonton, AB.

Olness, A. 1984. Re: nitrogen mineralization potentials, N_0 and correlations with maize response. Agron. J. 76:171–172.

Oyanedel, C., and J. Rodriguez S. 1977. Estimation of N mineralization in soils. Cienc. Invest. Agrar. 4: 33–44.

Paul, E.A., and N.G. Juma. 1981. Mineralization and immobilization of soil nitrogen by microorganisms. Ecol. Bull. (Stockholm) 33:179–194.

Paustian, K., and T.A. Bonde. 1987. Interpreting incubation data on nitrogen mineralization from soil organic matter. p. 101–112. *In* J.H. Cooley (ed.) Soil organic matter dynamics and soil productivity. Proc. from an INTECOL Workshop, Flen, Sweden. 4–6 June 1986. INTECOL Bull. 15. Int. Assoc. for Ecol., Athens, GA.

Power, J.F., J.W. Doran, and W.W. Wilhelm. 1986. Uptake of nitrogen from soil, fertilizer and crop residues by no-till corn and soybean. Soil Sci. Soc. Am. J. 50:137–142.

Rennie, D.A., C.A. Campbell, and T.L. Roberts. 1993. Impact of macronutrients on crop responses and environmental sustainability on the Canadian prairies — a review. Can. Soc. Soil Sci., Ottawa, ON.

Seligman, N.C., and H. van Keulen. 1981. PAPRAN. A simulation model of annual pasture production limited by rainfall and nitrogen. p. 192–22.1 *In* M.J. Frissel and J.A. van Veen (ed.) Simulation of nitrogen behaviour of soil–plant systems. PUDOC, Wageningen, the Netherlands.

Smith, S.J., L.B. Young, and G.E. Miller. 1977. Evaluation of soil nitrogen mineralization potentials under modified field conditions. Soil Sci. Soc. Am. J. 42:74–76.

Stanford, G. 1973. Rationale for optimum nitrogen fertilization in corn production. J. Environ. Qual. 2:159–166.

Stanford, G. 1982. Assessment of soil nitrogen availability. p. 651–688. *In* F.J. Stevenson (ed.) Nitrogen in agricultural soils. Agron. Monogr. 22. ASA, CSSA, SSSA, Madison, WI.

Stanford, G., and E. Epstein. 1974. Nitrogen mineralization-water relations in soils. Soil Sci. Soc. Am. Proc. 38:103–107.

Stanford, G., M.H. Frere, and D.H. Schwaninger. 1973. Temperature coefficient of soil nitrogen mineralization. Soil Sci. 115:321–323.

Stanford, G., and S.J. Smith. 1972. Nitrogen mineralization potentials of soils. Soil Sci. Soc. Am. Proc. 36:465–472.

Tietje, O., and M. Tapkenhinrichs. 1993. Evaluation of pedo-transfer functions. Soil Sci. Soc. Am. J. 57:1088–1095.

6 Current Phosphorus Availability Indices: Characteristics and Shortcomings[1]

F. R. Cox

North Carolina State University
Raleigh, North Carolina

The majority of soil tests for P currently employed in the USA use one of four extractants. The Bray-1 was developed in the 1940s (Bray & Kurtz, 1945) and the Mehlich-1 and Olsen in the 1950s (Mehlich, 1953; Olsen et al., 1954). Later, in the 1980s, the Mehlich-3, with characteristics similar to the Bray-1 for P, was developed (Mehlich, 1984). These extractants were designed to rapidly assess which soils should respond to P fertilization. Researchers have also desired that the concentration of P extracted would indicate something about the rate of P that should be applied, at least for a particular group of soils (Evans, 1987).

Over the years, the P concentrations extracted with these solutions have been compared extensively, correlated with P uptake and calibrated with yield response (Fixen & Grove, 1990). When kept within the obvious limitation of not using an acid extractant on a calcareous soil, there has generally been good agreement among extractants and each seems quite effective, at least for a limited range of soil conditions. Some researchers have altered the soil/solution ratio to fit local needs, and this should be noted when comparing results.

SHORTCOMINGS WITH CURRENT INDICES

Although these extractants have a long history of use and have been extensively researched, there are still shortcomings when they are used in a soil testing program. Those shortcomings, however, should not be confused with other, more obvious faults in the total soil testing program that includes phases from sampling all the way to the final fertilizer recommendation. It is very difficult to separate the shortcomings of current extractants from the numerous other problems that exist within a soil testing program.

Evidence of problems in the current soil testing program for P is apparent from the variation in the final outcome, the P fertilizer recommendation. Olson et al. (1987) reviewed the effects of fertilizer recommendations made by five laboratories on the yield of corn (*Zea mays* L.) grown at several locations in Nebraska.

---

[1]The research reported in this publication was funded by the North Carolina Agric. Res. Service.

Yields were similar among the laboratory recommendations, but the cost of fertilizer, including P, was markedly higher for four commercial laboratories relative to the University of Nebraska laboratory. For example, at the Mead site the commercial laboratories recommended 25 to 30 kg P ha^{-1} and the University laboratory only 15 kg P ha^{-1} (Fig. 6–1). The variation in cost is not a direct shortcoming of the extractant per se, as Olson et al. (1987) attributed the differences in fertilizer rates largely to the soil test interpretation philosophy; the build-up-maintenance concept of the commercial laboratories vs. the sufficiency concept of the University laboratory. It should be noted that the lower rate suggested by the University kept the soil test level well above the sufficiency level, the cut-off level above which no fertilizer is recommended.

Although almost all university laboratories use the sufficiency concept of soil test interpretation for P, there is still a marked range in P recommendations. Among five states in the Southern Region that use the Mehlich-1 extractant the extractable P concentration beyond which no fertilizer is recommended for corn ranges from 56 to 134 kg ha^{-1} (50–120 lb acre^{-1}; Fig. 6–2). Similarly, among four states in the Southern Region that use the Mehlich-3, the extractable P beyond which no fertilizer is recommended for corn ranges from 30 to 80 mg L^{-1} (Fig. 6–3). As most of these states adhere to the sufficiency concept of soil test interpretation, there has to be other reasons for the wide discrepancy in the cut-off point for fertilization. One of the reasons is undoubtedly the model used for calibration. Mallarino and Blackmer (1992) recently reiterated the importance of this factor. Another, perhaps more likely reason for the discrepancy, though, is differences in philosophy of interpretation, even within a method such as the sufficiency concept. Strategies vary, such as whether fertilization should cease at the

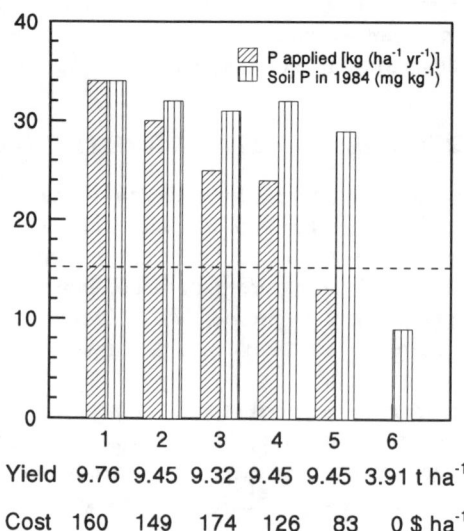

Fig. 6–1. Average corn yield, fertilizer cost, annual rate of P applied, and residual P after using recommendations of four commercial laboratories (1–4), the University of Nebraska Laboratory (5), and a check (6) after 12 yr at the Mead Field Laboratory in Nebraska (Olson et al., 1987).

CURRENT PHOSPHORUS AVAILABILITY INDICES

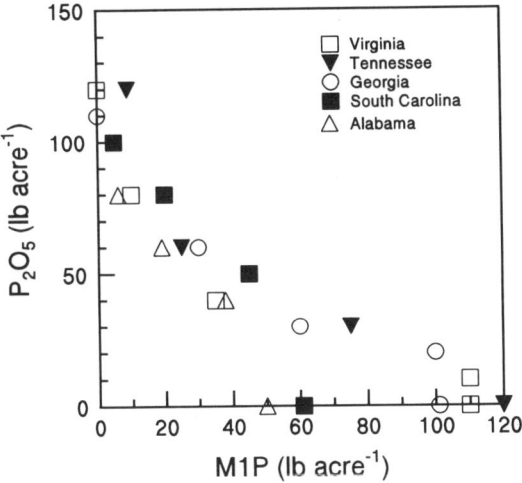

Fig. 6–2. Phosphorus recommendations for corn in five states in the Southern Region that use the Mehlich-1 extractant. For three of the states the soils are restricted to those which are sandy or low cation-exchange capacity.

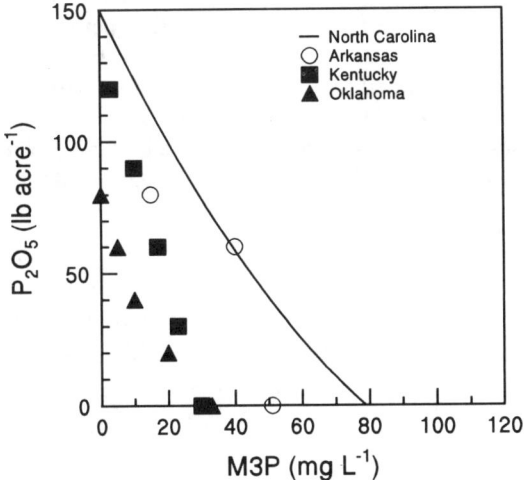

Fig. 6–3. Phosphorus recommendations for corn in four states in the Southern Region that use the Mehlich-3 extractant. The recommendations shown from Arkansas are for sandy loam soils.

sufficiency or critical level, or whether fertilization should be continued beyond that point. Reasons for the latter are likely centered on knowledge that there is no exact critical level, but a range that occurs due to biological conditions (Cox, 1992). This topic deserves much further attention, but such differences are not usually due to the selection of an extractant and the use of current P availability indices.

PROBLEM SOLUTION

Grouping Soils

In the data from the Southern Region (Fig. 6–2 and 6–3), there had been some grouping of soils. The groups in that area are based on exchange capacity, region (Coastal Plain vs. Piedmont), or texture, so the data were selected to represent the sandier, lower cation-exchange capacity (CEC) soils. Current P soil tests are not applicable across a broad range of soil conditions. A properly calibrated soil test should identify (i) the degree of deficiency or sufficiency of an element, and (ii) how much of the element should be applied if it is deficient (Evans, 1987). With current extractants, the concentration of extractable P alone cannot fulfill these criteria across a broad range of soil conditions because of differences in P buffer capacity.

Several states in the southeast are divided into regions for P soil test interpretation. One common method is to split out the soils in the Piedmont from those in the remainder of the state, especially the Coastal Plain where much of the agriculture is concentrated. The current P recommendations for corn grown in the Coastal Plain and Piedmont regions of South Carolina are shown in Fig. 6–4. Kamprath (1978) also noted that the critical level of P is lower in the clayey Piedmont soils than in the sandy Coastal Plain soils.

Another example of the shortcomings of the current P soil tests was given by Wendt and Corey (1981), who found no correlation between P uptake by corn grown in the greenhouse and Bray-1 P among several soils from Wisconsin (Table 6–1). If they determined the relationship with individual soils or soils of a single textural class, the r value was significant. An awareness of differences in both topsoil and subsoil characteristics has led to a grouping of soils in Wisconsin for the P soil test interpretation (S.M. Combs, 1993, personal communication).

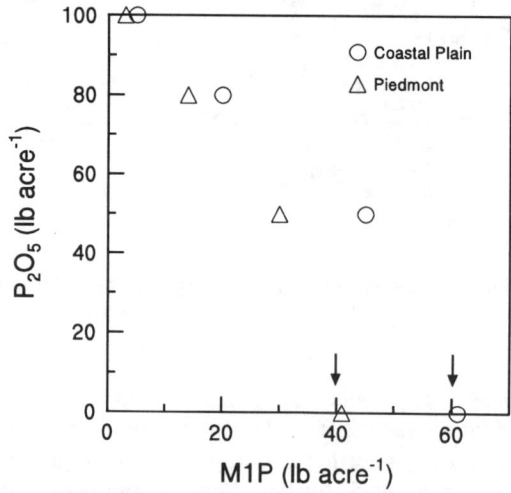

Fig. 6–4. Phosphorus recommendations for corn grown in the Coastal Plain and Piedmont of South Carolina in relation to Mehlich-1 extractable P.

Table 6–1. Correlation between Bray-1 P and P uptake by greenhouse corn (Wendt & Corey, 1981).

Soil†	n	r
Plainfield s	19	0.86**
Plano sil	15	0.97**
Fayette sil	15	0.91**
Withee sil	18	0.49*
Kewanee sicl + Hibbing sicl	18	0.71**
Total soils	90	0.13

*,** Significant at the 0.05 and 0.01 probability levels, respectively.
† s, sand; sil, silt loam; sicl, silty clay loam.

Olson (1987) also considered that subsoil P concentration may play an important role in the P soil test interpretation, but no assessment was given.

Quantifying Effects of Other Properties

Subsoil P is not a concern in the Southeast as the level is almost nil in the Ultisols of the region, but soils are often grouped by topsoil characteristics. Rather than just grouping soils for interpretation, however, it would be better to assess differences based on the continuum of soil properties as they exist in the field. We have been able to evaluate critical levels and quantities of P required to maintain a critical level with long-term residual studies on a Georgeville soil (clayey, kaolinitic, thermic Typic Hapludult) in the Piedmont and a Portsmouth soil (fine-loamy over sandy or sandy-skeletal, mixed, thermic Typic Umbraquult) in the Coastal Plain (Cox & Lins, 1984). In these studies, the effects of P rate and time had been determined (Cox et al., 1981) with the following expression:

$$X = X_{eq} + [(X_o + b_1 F + b_2 F_2) - X_{eq}] \exp(-kT) \quad [1]$$

In this expression, P applied in fertilizer (F) increases the soil test level (X), which then decreases in time (T), eventually approaching a minimum level (X_{eq}) asymptotically. An example of this relationship for the Portsmouth soil is shown in Fig. 6–5. Initial fertilizer treatments were no P and 171 kg ha^{-1}. The expression was fit with the data where no additional P was applied, but was also applicable where 20 kg P ha^{-1} was applied annually.

Yield was expressed with an exponential function:

$$Y = A - B \exp(-CX) \quad [2]$$

where Y is the yield in kg ha^{-1} and X is extractable P, also in kg ha^{-1}. An example of the response for corn grown on a Portsmouth soil in 1982 is shown in Fig. 6–6 (Cox & Lins, 1984).

Equations [1] and [2] were combined to make yield a function of soil and fertilizer P for a 1-yr period. Net income was calculated by multiplying the price of corn times the yield and subtracting the product of fertilizer amount and cost.

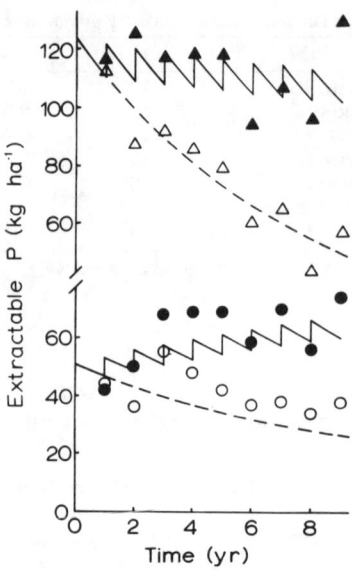

Fig. 6–5. Effect of initial P applications of 0 and 171 kg ha^{-1} and annual applications of 0 and 20 kg ha^{-1} on Mehlich-1 extractable P with time on a Portsmouth soil (Cox et al., 1981).

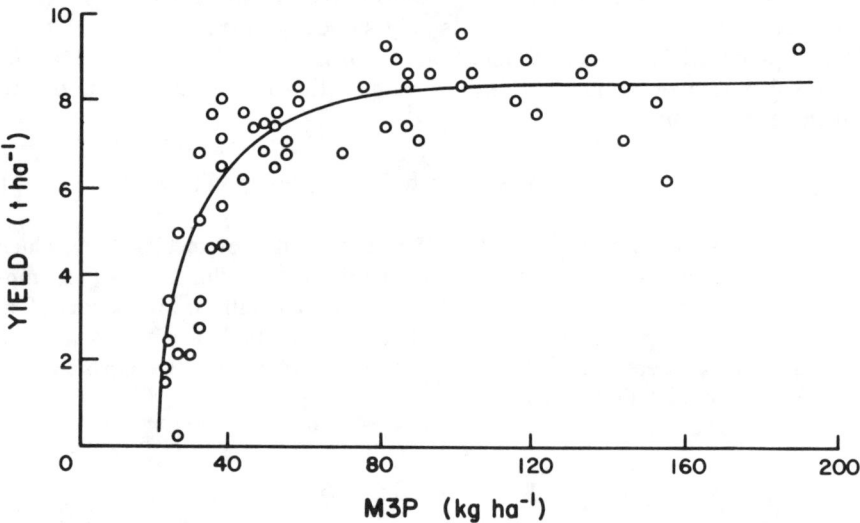

Fig. 6–6. Yield of corn grown on a Portsmouth soil in North Carolina in 1982 in relation to Mehlich-3 extractable P (Cox & Lins, 1984).

Setting the derivative of this expression to zero gave conditions for maximum net return. The solution at various soil test P levels gave a number of rates of P required for maximum net return. When this was done for the Portsmouth and Georgeville soils, the results were markedly different (Fig. 6–7). The relationship between fertilizer P required to maximize net income and soil test P was linear in each case, but no fertilizer would be suggested on the clayey Georgeville beyond ~20 kg M3P ha^{-1}, whereas fertilization should continue on the sandy Portsmouth to ~80 kg M3P ha^{-1}.

The difference in clay content is probably the main factor contributing to the variation in P fertilizer suggested in relation to soil test P. A multiple regression equation was formulated to include both soil test P and clay content. The quadratic term for clay was necessary and the final form of the equation was:

$$F = 107 - 0.7(M3P) + 0.072(clay)^2 - 0.0073(M3P)(clay)^2 \qquad [3]$$

This equation gives the P rate (F) for maximum economic yield at any combination of extractable P and clay content. It is based, however, on very limited information; only two sites and only 1 yr at each.

This concept also was used for soybean [*Glycine max* (L.) Merr.] grown on three soils for several years in Brazil (Lins et al., 1985). Yield response was calculated during 4 yr, which gave a good estimate of the most economic P rate for each soil. The relationship between the most economic P rate required and both Mehlich-1 extractable P and clay content indicated that at high clay content, the critical level would be lower, but more P would be required per unit of soil test P (Fig. 6–8).

In a continuation of the work in Brazil (Lins & Cox, 1989) several extractants were compared and factors other than clay content were used to predict the

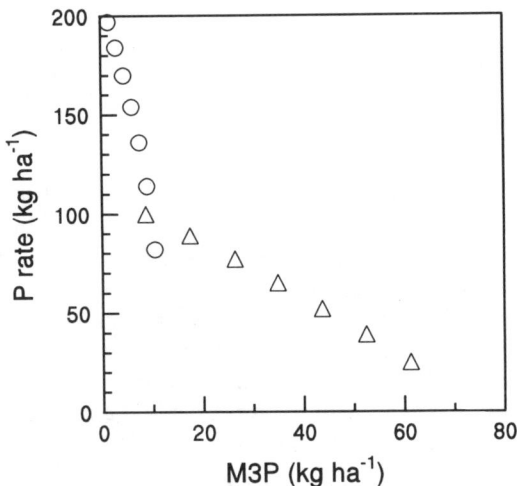

Fig. 6–7. Phosphorus recommendations for optimum corn production on a clayey soil (Georgeville, circle symbols) and a sandy soil (Portsmouth, triangle symbols) in North Carolina (Cox & Lins, 1984).

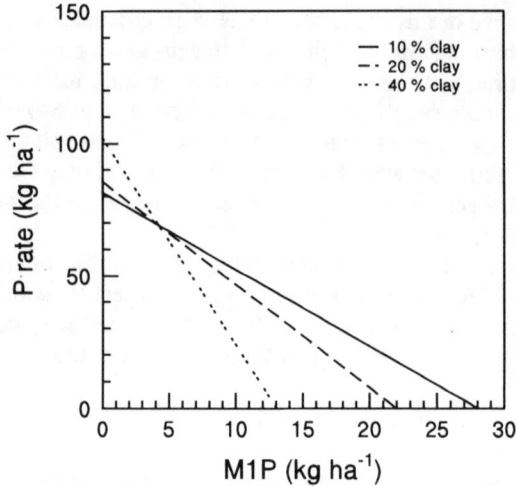

Fig. 6–8. Phosphorus recommendations for optimum soybean production on soils varying in clay content in Brazil (Lins et al., 1985).

Table 6–2. Relationship between optimum P rate and extractable P with or without other soil properties (Lins & Cox, 1989).

Property	Mehlich-3	Bray-1	Resin
		R^2	
Extractable P	0.67	0.57	0.62
Extractable P +			
Clay	0.80	0.78	0.95
Clay^2	0.84	0.83	–
Surface area	0.82	0.81	0.95
Surface area^2	0.84	0.84	–
P adsorption maximum	0.76	0.69	0.94
Buffer coefficient McLean	0.82	0.73	0.94

optimum P fertilizer rate. When used alone, the Mehlich-3, Bray-1, and a resin method (Raij & Quaggio, 1983) gave a coefficient of determination of ≈0.6 with optimum P rate (Table 6–2). With the Mehlich-3 and Bray-1, the squared terms of clay or surface area increased the coefficient to around 0.8, a somewhat higher value than achieved with P adsorption maximum or a P buffering coefficient. When the resin P was combined with any of the factors, the coefficient of determination increased to >0.9, indicating the promise of the method. It should be noted, however, that even with the resin method another factor such as clay content must be included.

Limitations

Several problems have been encountered during the course of these studies. The increase in soil test P may or may not be linear with the rate of P fertilizer applied; however, nonlinear functions are more difficult to use to determine the effects of future applications of P. Second, there is a limited amount of calibration data available in this form, so it would be difficult to confirm the best model to describe the effect of the clay term, although the quadratic seems best for most extractants. And, finally, other researchers and soil testers would like to apply local data and be able to compare effects more directly. Splitting the interpretation phase into determining the critical level and calculating the rate of P to be suggested would clarify the presentation.

Limitations of using clay content were also shown during the course of these studies. This approach seemed effective on soils with predominantly kaolinitic clays, but it was less effective on soils with high amounts of gibbsite. Other soils with high Al, such as some with amorphous clays, and soils with very high free $CaCO_3$, would also react differently. Such soils may have more precipitation of P rather than sorption, and on such soils sidebanding of P may be beneficial or even required. Perhaps the greatest limitation, however, is that other data cannot be compared or incorporated readily; it is just too complex to be practical.

Simpler Approaches

Long-term P studies are readily used to determine the rate of P required to change the soil test P level by a unit. We have conducted additional studies in North Carolina to add to the database of work done in our state and in Brazil. The change in Mehlich-3 P 1 yr after applying a unit of fertilizer P is an exponential function of the clay content of the soil (Fig. 6–9). The overall shape of this curve is very similar to that found in South Africa by Johnston et al. (1991) with the Bray-1 extractant after a period of incubation (Fig. 6–10). Thus, given the clay content of the soil, one can calculate the amount of fertilizer required to increase the soil test to the critical level.

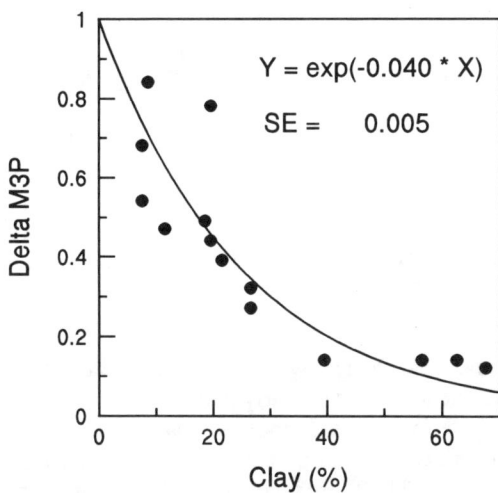

Fig. 6–9. The change in Mehlich-3 extractable P a year after applying a unit of P to soils of varying clay content (Cox, 1992).

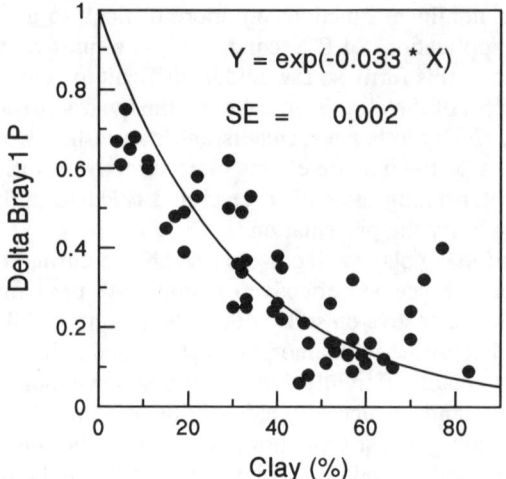

Fig. 6–10. The change in Bray-1 P per unit of P applied in relation to clay content for soils from several orders in South Africa (data from Johnston et al., 1991).

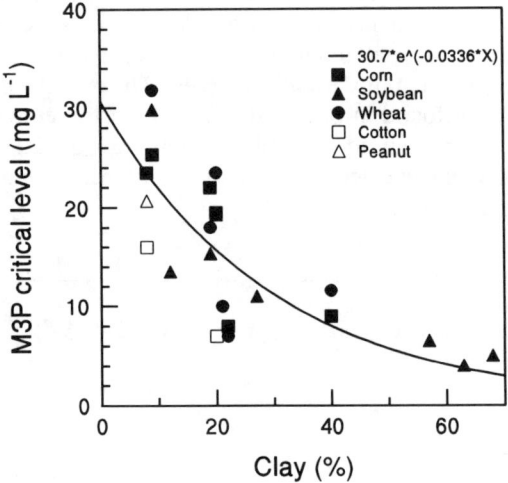

Fig. 6–11. Observed changes in the critical level of P in soil in relation to clay content when growing several crops in North Carolina.

But the critical level also changes with clay content. The critical level is more difficult to determine as one has to have clearly deficient conditions under good enough management and climate for clear definition. We have amassed a few estimates over several crops and the critical level also decreases with increasing clay content in an exponential manner (Fig. 6–11). There is considerable variability in the estimates, and there is no clear distinction among crops, but the relationship between critical level and clay content is clear. The P buffer capacity does affect the critical level.

The relationships shown in Fig. 6–9 to 6–11 indicate the importance of buffer capacity, as estimated by clay content, in the interpretation of current soil test indices. Both the critical level and the rate of fertilization required to be sufficient for a crop are dependent on the texture of the soil, and using these relationships would improve P fertilizer recommendations. Surface textures are often known from farm maps or can be estimated from county maps. In the Southeast, increased clay is closely associated with the color of the soil (Cox & Espejo, 1990), so clay could be estimated directly in any soil testing laboratory. However it is done, inclusion of clay in the interpretation of P will overcome one of the shortcomings of current soil test indices and improve P recommendations.

ACCOUNTING FOR DIFFERENT CRITICAL LEVELS

In reality, there will always be ambiguity associated with the critical level because it is associated with crop and soil biological factors that are responding to environmental conditions. In an evaluation of the critical level of P for corn grown on a sandy Ultisol in North Carolina over a period of years, there was marked variability among years (Table 6–3) (Cox, 1992). Critical levels found with an exponential model were similar at 95% maximum yield and when net return was calculated for corn, but <95% should have been used for soybean and wheat (*Triticum aestivum* L.). Critical levels with a linear response and plateau model were lower than those with the exponential model for all crops, but the

Table 6–3. Critical levels of Mehlich-3 extractable P for three crops determined with two mathematical functions (Cox, 1992).

	P critical level		
	Exponential function		
Year	95%	Economic[†]	Linear plateau
	mg L^{-1}		
	Corn		
1982	31	33	22
1984	46	43	33
1986	41	42	24
1988	54	50	33
1990	23	23	18
X	39	28	26
	Soybean		
1983	73	50	41
1985	36	30	23
1987	37	29	24
1989	53	35	31
X	50	36	30
	Wheat		
1985	82	47	39
1987	55	31	34
1989	55	38	33
X	64	39	35
SD[‡]	15	10	7

[†] Values used ($ kg^{-1}) : P = 1.27; corn = 0.1083; soybean = 0.2205; and wheat = 0.147.
[‡] Standard deviation.

standard deviation was about one-fourth of the mean with each approach, indicating considerable variation among years. The Mehlich-3 P critical level for corn with the linear response and plateau model on this soil with <10% clay was 26 mg L^{-1}.

The percentage of maximum yield used to divide responsive and nonresponsive soils with the exponential function should be selected from that giving the highest net return. Mallarino and Blackmer (1992) determined the response of corn to P at 25 sites, most of which were Mollisols, in Iowa. There were positive responses on six of the sites. They calculated the critical level and net return when applying 25 kg P ha^{-1}, a rate that usually gave maximum yield, with several models based on the relative yield. The models included exponential and quadratic response and plateau ones at several percentages of maximum yield, linear response and plateau, and the Cate-Nelson (Cate & Nelson, 1971) split (Table 6–4). For corn, dividing responsive and nonresponsive sites at 95% of maximum yield gave the highest net return for the exponential and quadratic response and plateau models. The linear plateau gave nearly the same return and the Cate-Nelson gave the highest net return. The authors also tried other extractants and other means of expressing yield response, but concluded that getting a clear definition of the critical level was paramount to achieving maximum economic yield. The Mehlich-3 P critical level for corn with the linear response and plateau model for these soils, which probably contained about 25% clay, was 14 mg kg^{-1}, or ~14 mg L^{-1}.

The critical level found on the sandy Ultisol in North Carolina was twice that found on the Mollisols in Iowa, as noted above. Much of this difference should be accounted for with the P sorption capacity or P buffer capacity of the topsoil, as identified by the clay content. According to the relationship shown in Fig. 6–11, the critical level would be 22 mg L^{-1} for a soil with 10% clay and 13 mg L^{-1} for a soil with 25% clay. Recognition of this effect will help account for differences reported in critical levels from various regions.

Table 6–4. Critical levels of M3P, net return, and number of fields fertilized of 25 sites in Iowa evaluated with several models at various percentage maximum yields of corn (Mallarino & Blackmer, 1992).

Model	Maximum yield	Critical M3P	Net return	Fields fertilized
	%	mg kg^{-1}	$	
Exponential	99	18	106	14
	95	11	318	4
	90	8	200	2
Quadratic	100	18	275	9
plateau	99	15	275	9
	95	11	318	4
	90	8	200	2
Linear plateau	100	14	294	7
Cate-Nelson	100	12	395	5

REFERENCES

Bray, R.H., and L.T. Kurtz. 1945. Determination of total, organic and available forms of phosphorus in soils. Soil Sci. 59:39–45.

Cate, R.B., Jr., and L.A. Nelson. 1971. A simple statistical procedure for partitioning soil test correlation data into two classes. Soil Sci. Soc. Am. Proc. 35:658–660.

Cox, F.R. 1992. Range in soil phosphorus critical levels with time. Soil Sci. Soc. Am. J. 56:1504–1509.

Cox, F.R., and R. Espejo. 1990. Readily measurable soil properties that affect the phosphorus fertilizer requirement of Ultisols. Commun. Soil Sci. Plant Anal. 21:2079–2088.

Cox, F.R., E.J. Kamprath, and R.E. McCollum. 1981. A descriptive model of soil test nutrient levels following fertilization. Soil Sci. Soc. Am. J. 45:529–532.

Cox, F.R., and I.D.G. Lins. 1984. A phosphorus soil test interpretation for corn grown on acid soils varying in crystalline clay content. Commun. Soil Sci. Plant Anal. 15:1481–1491.

Evans, C.E. 1987. Soil test calibration. p. 23–29. In J.R. Brown (ed.) Soil testing: Sampling, correlation, calibration, and interpretation. SSSA Spec. Publ. 21. SSSA, Madison, WI.

Fixen, P.E., and J.H. Grove. 1990. Testing soils for phosphorus. p. 141–180. In R.L. Westerman (ed.) Soil testing and plant analysis. 3rd ed. SSSA Book Ser. 3. SSSA, Madison, WI.

Johnston, M.A., N. Miles, and G.R. Thibaud. 1991. Quantities of phosphorus fertilizer required to raise the soil test value. S. Afr. J. Plant Soil 8:17–21.

Kamprath, E.J. 1978. The role of soil chemistry in the diagnosis of nutrient disorders in tropical situations. p. 313–327. In C.S. Andrew and E.J. Kamprath (ed.) Mineral nutrition of legumes in tropical and subtropical soils. CSIRO, Canberra, Australia.

Lins, I.D.G, and F.R. Cox. 1989. Effect of extractant and selected soil properties on predicting the optimum phosphorus fertilizer rate for growing soybeans under field conditions. Commun. Soil Sci. Plant Anal. 20:319–333.

Lins, I.D.G., F.R. Cox, and J.J. Nicholaides, III. 1985. Optimizing phosphorus fertilization rates for soybeans grown on Oxisols and associated Entisols. Soil Sci. Soc. Am. J. 49:1457–1460.

Mallarino, A.P., and A.M. Blackmer. 1992. Comparison of methods for determining critical concentrations of soil test phosphorus for corn. Agron. J. 84:850–856.

Mehlich, A. 1953. Determination of P, Ca, Mg, K, Na and NH_4. North Carolina Soil Testing Div. Mimeo, Raleigh.

Mehlich, A. 1984. Mehlich-3 soil test extractant: A modification of Mehlich-2 extractant. Commun. Soil Sci. Plant Anal. 15:1409–1416.

Olsen, S.R., C.V. Cole, F.S. Watanabe, and L.A. Dean. 1954. Estimation of available phosphorus in soils by extraction with sodium bicarbonate. USDA Circ. 939. U.S. Gov. Print. Office, Washington, DC.

Olson, R.A., F.N. Anderson, K.D. Frank, P.H. Grabouski, G.W. Rehm, and C.A. Shapiro. 1987. Soil testing interpretations: Sufficiency vs. build-up and maintenance. p. 41–52. In J.R. Brown (ed.) Soil testing: Sampling, correlation, calibration, and interpretation. SSSA Spec. Publ. 21. SSSA, Madison, WI.

Raij, B.van, and J.A. Quaggio. 1983. Metodos de analise de solo para fins de fertilidade. Boletim Tecnico 81. Instituto Agronomico, Campinas, Brazil.

Wendt, R.C., and R.B. Corey. 1981. Available P determination by equilibration with dilute $SrCl_2$. Commun. Soil Sci. Plant Anal. 12:557–568.

7 Innovative Soil Phosphorus Availability Indices: Assessing Inorganic Phosphorus

Andrew N. Sharpley

National Agricultural Water Quality Laboratory
Durant, Oklahoma

J. T. Sims

University of Delaware
Newark, Deleware

Gary M. Pierzynski

Kansas State University
Manhattan, Kansas

Sustainable crop production depends on the maintenance of adequate plant nutrients in the soil root zone. Nutrients most frequently required as external inputs are N, P, and K. The amounts of these nutrient inputs required for optimum crop yields have been determined by soil tests for plant available levels in the root zone and, to a lesser extent in plant tissue, for over a century. Recent research and advances in analytical and computing techniques, however, have led to more sophisticated and innovative indices quantifying nutrient input requirements. We focus here on advances in soil testing for inorganic P; issues concerning organic P, N, and K testing are covered in separate chapters of this special publication.

Soil testing is essential to accurate, profitable fertilizer recommendations. While soil testing does not directly tell us how much fertilizer P will be required for optimum crop yields, interpretation of soil analyses will guide producers as to the amount of supplemental P required. Interpretation of soil test results and the development of recommended rates of P application should be based on research data (public and private) that relates concentrations of soil test P (STP) to crop response. Recommendations should be refined regionally to obtain flexible yet effective soil testing-recommendation procedures.

Environmental implications of the percolation of fertilizer N through soil profiles as NO_3–N and the transport of P in surface runoff from agricultural land, have emphasized the importance of more closely balancing nutrient inputs with crop needs. As a result, there has been a need in the last decade to develop soil N

Copyright © 1994 Soil Science Society of America, 677 S. Segoe Rd., Madison, WI 53711, USA.
Soil Testing: Prospects for Improving Nutrient Recommendations, SSSA Special Publication 40.

and P tests that reflect both agronomic and environmental aspects of nutrient management. In areas of intensive crop and livestock production, soil P management is often of prime concern in developing sustainable management systems. This is a result of the imbalance between N and P inputs and crop removal, particularly where animal manures are used. For example, the ratio for N/P for several animal manures (Gilbertson et al., 1979) is 2, while crop removal of N/P for major grain and hay crops ranges from 8 to 22. Thus, soil P has accumulated to levels exceeding those required for optimum crop yields (Pierzynski et al., 1990; Sharpley et al., 1994; Sims, 1992). As a result, there has been an increased awareness of soil P fertility, along with the need for reliable, versatile, and simple soil P tests that identify not only soils deficient but those that are excessive in P as well.

For P, these needs have centered around questions concerning the development of innovative soil extraction and interpretation approaches for the recommendation of P applications as fertilizers and manures. With the increased adoption of conservation tillage systems and more precise subsurface placement and banding of fertilizer in the soil, are current soil sampling procedures adequate? Has the use of multi-element extractants, such as Mehlich 3 (Mehlich, 1984), affected the estimation of soil P availability? Can P-sink approaches, such as resin and Fe-oxide materials, provide practical as well as theoretical improvements to soil P testing? Can current soil tests estimating plant P availability, be used as environmental tests, to estimate the potential bioavailability of soil P in aquatic ecosystems, and if not, what tests would be more appropriate? And finally, as soils and agricultural management systems vary regionally, should STP be based on a regional rather than individual state basis?

This chapter attempts to answer these questions and presents recent developments in the estimation of soil P availability and their implications to the recommendation process for P inputs to agronomically, economically, and environmentally sound agricultural systems.

SOIL SAMPLING TECHNIQUES

Soil testing success depends on collection of a representative soil sample, in addition to accurate laboratory analyses. Numerous soil samples, from the appropriate soil depth, need to be collected from representative areas of the field and composited. When nutrients are applied uniformly (e.g., broadcast), a field should be divided, avoiding odd or problem areas, into about 8 to 16 ha lots (40 ha lots may be more practical, but less desirable) and 15 to 20 subsamples collected from each lot for an adequate composite sample. Different or special problem areas should be sampled separately. Clearly, a nonrepresentative sample is misleading and may be worse than no sample at all (Kitchen et al., 1990).

Collecting a representative soil sample is still one of the main limitations to reliable fertilizer recommendations (Ellis & Olson, 1986). Simply, the quality of a soil test result is only as good as the quality of the soil sample analyzed. For example, the disparity between the amount of soil in a field and that analyzed is clearly demonstrated by the simple calculation that a 1 ha field would have 2600 Mg of soil in a 0- to 20-cm depth, assuming a 1.3 Mg m^{-3} soil bulk density. Most soil P analyses require only 1 to 5 g of soil. Ellis and Olson (1986) proposed detailed protocol for collecting representative soil samples taking into account sampling equipment, uniformity of soil characteristics, field topography, previous fertilizer placement, and soil management.

Agronomic Sampling

The spatial distribution of P caused by fertilizer P banding, manure application, and reduced tillage, will present sampling problems for the determination of subsequent fertilizer P requirements. For example, if location of the fertilizer band is known, what portion of samples should be collected on and off the band and, if the band's location is not known, is a random sampling strategy adequate? Collection of 15 (Shapiro, 1988; Ward & Leikam, 1986) to 30 random samples (Hooker, 1976) has been reported to adequately reflect P availability in fields where P bands exist.

Sampling strategies for minimum and no-till conditions would be similar to conventional tillage situations when P has been broadcast applied, except for sampling depth. Although not widely adopted, it is suggested conservation tilled soils be sampled to a shallower depth (0 to 5 or 10 cm) compared with conventional tillage (0–20 cm; James & Wells, 1990; Whitney, 1982). The sampling strategies described by Kitchen et al. (1990), however, should be followed for reduced- and no-till situations where P has been banded. When location of the P bands are known, sampling involves one core in the band for every 20 out of the band, when band spacing is 76 cm, and a ratio of 1-in to 8-out when band spacing is 30 cm (Kitchen et al., 1990). When band location is unknown, paired sampling (~10) is recommended to reduce soil test P variability over completely random sampling. In this process, a random core is first collected followed by a second core 50% of the band-spacing distance from the first sample, perpendicular to the band direction.

Environmental Sampling

Soil sampling protocol for environmental concerns need to be reevaluated since the primary mechanism for P transport from most agricultural soils is by surface runoff and erosion. Although most soil samples submitted to soil testing laboratories are obtained from 0 to 20 cm, the zone of interaction of runoff waters with most soils is normally <5 cm. Consequently, environmental soil sampling should reflect this shallower depth of soil influencing runoff P. Hence, environmental soil samples should in general, be taken from no deeper than 5 cm. This protocol is compatible with sampling of no-till fields, currently recommended by extension specialists in several states, where the traditional 0- to 20-cm depth is split into two or three increments. Thus, on soils identified as vulnerable to P loss in runoff, the surface increment could be analyzed for environmental interpretation and all increments integrated for agronomic interpretations.

Given the lack of mobility of P in soils, the soil surface can become highly enriched in P, particularly when long-term annual P applications are made and incorporation is rare. For example, decades of P fertilization at rates exceeding the amount removed by crops have resulted in widespread increases in STP levels. Combs and Burlington (1992) reported an average Bray 1 P content of 48 mg kg^{-1} in soils tested in Wisconsin in 1990, compared with 34 mg kg^{-1} in 1967. Coarse textured soils, reflecting their extensive use in vegetable production, had an even higher average of 72 mg kg^{-1} of STP. Conservation tillage can increase the STP content of surface soil, if P is broadcast without tillage incorporation. In a long-term tillage study, Griffith et al. (1977) demonstrated the typical stratification of STP level under no-till conditions. Within a few years, the surface 2.5 cm of soil under no till was six times higher than initial STP levels. In a similar study, Guertal et al. (1991) reported that STP (Bray 1 P) values in the upper 2 cm of long-term no-tilled soils were 200 and 290 mg kg^{-1}; values for STP at the 6- to 8-cm depth of the same soils were 60 and 110 mg kg^{-1}.

Levels of STP in surface horizons of soil are also elevated by long-term application of manures and wastes (Sharpley et al., 1993; Sims, 1992). Application of dairy manure has contributed to 200 mg kg^{-1} STP (Bray 1 P) levels in Wisconsin (Motschall & Daniel, 1982) and Pierzynski et al. (1990) found levels of 613 mg kg^{-1} of STP (Bray 1 P) in Illinois as a result of sludge additions. Sharpley et al. (1991b) examined several Oklahoma soils receiving long-term application of poultry litter and found STP (Mehlich 3 P) levels of up to 279 mg kg^{-1}.

In summary, a modified sampling protocol is needed for environmental soil tests for P, because of the accumulation of P in the surface 5 cm of soil. Thus, a 0- to 20-cm sample would underestimate any accumulation of P in the surface 5 cm and not be a true reflection of the soils' potential to enrich P in runoff.

AGRONOMIC SOIL TESTING FOR PHOSPHORUS

Current Approaches

Soil tests for P were originally developed to correlate with the amount of plant-available P in topsoil horizons in order to rank soils in terms of the likelihood of yield response to P fertilization. Several excellent reviews of the agronomic principles and methodology of soil P testing are available (Fixen & Grove, 1990; Kamprath & Watson, 1980; Olsen & Sommers, 1982; Olson et al., 1987). Agronomic soil P tests now used in the USA are chemical solutions that contain dilute acids, fluoride, bicarbonate, or acetates (Table 7–1). Historically, the most common soil P tests have been Mehlich 1 (Mehlich, 1953), Bray 1 P (Bray & Kurtz, 1945), and Olsen-P (Olsen et al., 1954); however, a number of laboratories have converted to the Mehlich 3 (Mehlich, 1984) or AB-DTPA (Soltanpour & Schwab, 1977) soil tests in the past decade. Each of these solutions was designed to extract similar forms of soil P believed to be accessible to plant roots: soluble P, Ca-P, Al-P, Fe-P, and organic P. The mode of action of these extractants varies somewhat, but generally involves acid dissolution, anion exchange, cation complexation, or cation hydrolysis (Kamprath & Watson, 1980). In some cases (Bray 1 P and Olsen P), soil testing for P involves a separate extraction of a soil sample with a different solution than is used for K, Ca, Mg, or micronutrients (Cu, Fe, Mn, and Zn), while other soil tests (Mehlich 1, Mehlich 3, and AB-DTPA) are multi-element extractants.

It is unlikely that an extractant would exclusively measure a single pool of soil P, although some components of extractants are aimed at specific pools. For

Table 7-1. Current soil testing extractants used for P by states in four regoinal soil testing committees in the USA.†

Soil test extractant, original reference, and composition	States using extractant
Morgan (Morgan, 1941) (0.72 M NaOAc + 0.52 M CH_3COOH, pH 4.8)	MA, NY
Modified Morgan (McIntosh, 1969) (0.62 M NaOAc + 1.25 M CH_3COOH, pH 4.8)	CT, ME, NH, VT
Mehlich 1 (Mehilich, 1953) (0.0125 M H_2SO_4 + 0.05 M HCl)	AL, DE, FL, GA, MD, RI, SC, TN, VA, WV
Mehlich 3 (Mehlich, 1984) (0.2 M CH_3COOH + 0.25 M NH_4NO_3 + 0.015 M NH_4F + 0.013 M HNO_3 + 0.001 M EDTA)	AR, KY, NC, NJ, OK, PA
Bray & Kurtz no. 1 (Bray & Kurtz, 1945) (0.025 M HCl + 0.03 M NH_4F)	IA, IN, KS, LA, MI, MN, MO, NE, OH, SD, WI
Olsen-P (Olsen et al., 1954) (0.5 M $NaHCO_3$, pH 8.5)	MN, ND, NE, SD

†Regional soil testing committees include Northeast Coordinating Committtee on Soil Testing (NEC-67), North Central Soil Testing Committtee (NCR-13), Southern Regional Soil Testing Information Exchange Group (STSRIEG), and Mid-Atlantic Soil Testing and Plant Analysis Work Group (MASTPAWG). See text for information on states participating in each committee.

example, F of the Bray 1 extractant exchanges with Al-bound P, with the assumption that this Al-bound P contributes to the P quantity in acid soils. The success of any extractant to estimate plant available P depends on the appropriateness of the chemical used relative to soil properties and on the empirical relationship between the measured quantity of P and crop needs. Alternative approaches use P-sinks, such as anion exchange resins and Fe-oxide impregnated paper, to determine the quantity of soil P available to plants with negligible chemical extraction.

Calibration and Correlation of Soil Tests for Phosphorus

The accuracy of these soil tests in identifying P deficient soils has traditionally been evaluated in field and greenhouse studies. Two types of studies are used, *correlation*, where P uptake by plants is related to STP, usually in greenhouse experiments with soils of varying properties; or *calibration* experiments that measure the magnitude of crop yield increase obtained when a soil with a certain STP level is fertilized. Fixen and Grove (1990) summarized the results of dozens of field and greenhouse correlation and calibration studies and concluded that STP is "...best at predicting the probability of P response, poorer at predicting the magnitude of any response, and weakest at determining the exact rate of P needed for optimum economic performance in any given year or field." They also stated that "...nearly any soil test for P will fail if used on inappropriate soils." Innovations in soil testing for P, therefore, must include not only the development of new extracting solutions or methods, but a commitment to field or greenhouse studies that verify the accuracy of the new test for a range of soils.

Unfortunately, one of the greatest concerns about agronomic soil testing for P is the lack of ongoing calibration research that reflects changes in tillage practices, genetic improvement of crops, and fertilizer sources or application techniques. Of the 47 correlation studies cited by Fixen and Grove (1990), only three had been conducted since 1980. Inadequate funding from all sources (universities, agribusiness, and government) has been a major factor contributing to the failure of soil testing calibration research to keep pace with changes in production agriculture. This means that most studies conducted to evaluate innovations in soil testing for P will be conducted under the rather artificial conditions of a greenhouse because of the considerable expense associated with multiyear, multisite field calibration research.

A typical example of short-term correlation research was the greenhouse study of Bates (1990) that compared two relatively new soil test extractants for P (AB-DTPA and Mehlich 3) with three existing extractants (Bray 1, Bray 2, and Olsen) using 88 soils from Ontario, Canada. All extractants were well correlated with each other, in terms of extractable P (r^2 = 0.69–0.94), and with P uptake by corn (*Zea mays* L.; r^2 = 0.53–0.74), suggesting that conversion from one of the existing soil tests to either Mehlich 3 or AB-DTPA would be acceptable. Greenhouse studies such as this, however, bypass some of the more critical aspects of soil test development, such as the influences of climatic variations, crop rotations and subsoil effects on crop response to P. Because of this, they can hardly be viewed as the most desirable approach to innovations in soil testing, particularly since field calibration studies rarely achieve the same degree of predictive accuracy as greenhouse experiments. In one recent study, Beegle and Oravec (1990) evaluated the Mehlich 3 and Bray 1 extractants as soil tests for corn at 67 sites in Pennsylvania. The two soil tests were highly correlated with each other (r^2 = 0.98) and had similar critical values for P (Mehlich 3: 45 kg P ha^{-1} and Bray 1: 43 kg P ha^{-1}). The authors noted *wide scatter* in the data, however, and stated that "...the poor relationships between the soil test levels of any of the extractants and crop yield...make a strong case for a more mechanistic approach to soil testing."

Some research has also been conducted to assess the time required to deplete soils that are excessive in STP to levels where fertilization will once again be required. This has both agronomic and environmental implications, as discussed in more detail below. Because this depletion can often take 10 yr or more, most of these studies have also been conducted in the greenhouse (Adepoju et al., 1982; Aquino & Hanson, 1984; Bowman et al., 1978; McCallister et al., 1987; Novais & Kamprath, 1978). Extrapolation to field conditions, thus, requires assumptions about the relationship between P uptake in a greenhouse pot and crop P removal in the field. For example, Novais and Kamprath (1978) used regression equations between plant uptake and STP depletion obtained from a successive cropping greenhouse study, along with estimates of P removal by corn from an 8-yr field study, to rank North Carolina soils in terms of long-term P supplying capacity. Estimates of the time required to deplete STP from initial values ranging from 53 to 165 mg STP kg^{-1} to the critical STP level, based on the Mehlich 1 soil test, (≈20 mg STP kg^{-1}) ranged from 1 to 6 yr. McCollum (1991) reported, however, the results of a 30-yr study in North Carolina and estimated that, with no further addition of P, 16 to 18 yr of corn or soybean [*Glycine max* (L.) Merr.] production would be needed to reduce STP in a Portsmouth soil (fine-loamy over sandy or sandy-skeletal, mixed, thermic Typic Umbraquult) from 100 to 20 mg kg^{-1} Mehlich 1 P, the critical value at which the soil would again be

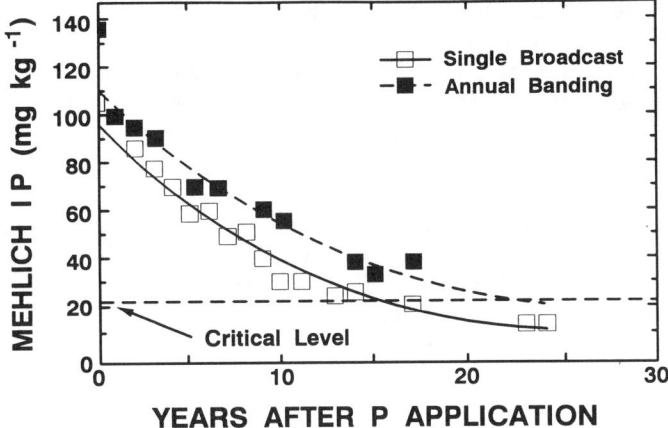

Fig. 7–1. Decrease in soil test (Mehlich 1, 0.05 M HCl + 0.0125 M H_2SO_4) extractable P in a Portsmouth soil cropped to corn–soybean rotation for 26 yr. Initial soil test P levels ($T = 0$) were established by a single broadcast application of 324 kg P ha^{-1} or eight annual banded applications of P at 60 kg P ha^{-1} (adapted from McCollum, 1991).

considered responsive to P applications (Fig. 7–1). This example again illustrates the problems in improving soil testing for P without adequate, long-term field calibration research.

Innovations in Agronomic Soil Testing

Innovative approaches to the laboratory assessment of the availability of soil P for agronomic purposes have been reported, but few have been implemented by soil testing laboratories. These innovations can be grouped into three categories: (i) new extracting solutions; (ii) new approaches to extract plant-available P from soils; and (iii) indexes or models that use STP in conjunction with other soil properties. Each of these are discussed below.

Research to develop more effective soil testing extractants for P has been conducted on a fairly regular basis for decades. Astonishingly, however, only two new soil test solutions, the Mehlich 3 and AB-DTPA, have received widespread adoption in the 40 plus years since the development of the Bray 1 (Bray & Kurtz, 1945), Mehlich 1 (Mehlich, 1953), Morgan (Morgan, 1941), and Olsen (Olsen et al., 1954) soil tests. Further, the major reason for the development of the two new soil tests (Mehlich 3 and AB-DTPA) was not to improve the predictive accuracy of the soil test for P, but to allow soil testing laboratories in the eastern (Mehlich 3) and western (AB-DTPA) USA to save time and expense by replacing two or more extractants for separate elements with one, multi-element soil test. Indeed, design of the new extractants was often meant to mimic the original two solutions, with perhaps some modification to improve the effectiveness at micronutrient soil testing. For example, the Mehlich 3 extractant, which often replaces three soil tests, the Bray 1 (for P), ammonium acetate (for K, Ca, and Mg), and EDTA (ethylenediaminetetraacetic acid; for Fe, Mn, Cu, and Zn), is an acidic extractant containing ammonium, fluoride, and EDTA, major components of the three tests it replaced.

This is not to say that Mehlich 3 and AB-DTPA are not equal in value to the P soil tests they replaced, for several studies have shown them to be so (Bates, 1990; Hanlon & Johnson, 1984; Labhsetwar & Soltanpour, 1985; Sims, 1989; Wolf & Baker, 1985). It does imply, however, either: (i) that decades of soil chemistry research have not resulted in soil testing solutions sufficiently superior to those currently in use to justify their inclusion in modern soil testing programs; or, perhaps more likely; (ii) new methods may not show much improvement in the field over current methods due to difficulties in obtaining representative soil samples and the uncontrollable effect of climate on plant growth; and (iii) that soil testing programs are very conservative in nature and reluctant to change extractants until adequate long-term, multiple site, multiple crop research and education has been conducted to clearly demonstrate the superiority of the new extractant over the old. Unfortunately, because of the funding constraints mentioned earlier, it is highly unlikely that these types of studies will be conducted; hence, it appears that, for now, innovation in the choice of extracting solutions for P will either proceed more or less on faith and intuition or not at all.

Another innovation in soil testing for P has been the development of new methods to extract P from soils. The goal in this case has been to develop a practical alternative, with a better theoretical basis, to the chemical extracting solutions now used by most routine soil testing laboratories. These methods can be grouped under P-sink approaches, described in the next section.

A third innovation in agronomic soil testing for P has been the use of indexes or models that combine STP with other soil properties related to P intensity and buffer capacity, to increase predictive accuracy. Because, for the most part, the more complex, computer-based models, such as EPIC (Erosion-Productivity-Impact-Calculator; Jones et al., 1984a,b) are applied at the watershed scale, this section focuses on P indexes that can be calculated from the results of a routine soil test and any information on soil type and crop management practices provided on the information sheet that accompanies a soil sample. Practically speaking, to be successfully used by a routine soil testing laboratory, any such index or model must rely upon easily measured or estimated soil properties. Preferably the index would contain parameters already measured (pH, organic matter, and cation-exchange capacity) by the laboratory.

Ideally, the information provided to the soil testing laboratory would contain accurate descriptions of the soil series from which the sample was collected; realistically, if the geographic location of the sample is known, it may be possible to make assumptions about soil type that will improve the interpretation of analytical results for STP. Geographic information systems that overlay soil series with land use may facilitate this process in the future.

One example of a simple P index was provided by the study of Bates (1990), mentioned earlier. Results of their greenhouse experiments showed that coefficients of determination (r^2) between STP and plant P uptake were increased from 0.54–0.74 (STP only) to 0.70–0.80 merely by inclusion of soil pH in a multiple regression equation. Lins et al. (1985), recognizing that the considerable variation in P sorption capacity of soils can markedly influence P availability, developed a soil test interpretation for P using multiple regression models based on STP and clay content. Using this approach they identified critical levels for Mehlich 1 P of 6, 15, and 23 mg STP kg^{-1} for soils with clay contents of 63, 27, and 12%, respectively. Clay content, however, is much more cumbersome and time-consuming to determine than pH; hence, to be successful on a practical scale the approach of Lins et al. (1985) would require some surrogate measure for percentage of clay. Examples might include hand estimates of textural class, volume weight (often obtained by laboratories measuring organic matter by loss on ignition), or even soil mapping unit from a soil survey manual.

PHOSPHORUS-SINK APPROACHES

The concept of exposing the soil to a P-sink has merit toward the goal of assessing bioavailable inorganic P (i.e., available to plants and algae) for both agronomic and environmental goals. Presumably, this would allow only P that was able to respond to such a sink to be measured, which is analogous to a root acting as a sink in the soil or to the concentration gradient that exists when a small quantity of sediment is placed in a large volume of water. The analogy of a root is not entirely accurate because root exudates and mycorrhizae fungi can alter P availability in the rhizosphere such that the root does not behave as a pure sink. Still, P-sinks are probably the closest manifestation of the root environment that is available. Some authors assume that the sink maintains extremely low P concentrations in the aqueous media employed and can be considered an infinite P-sink in the sense that P release by the soil is clearly the rate-limiting step (Sibbesen, 1978; van der Zee et al., 1987; Yli-Halla, 1990). For anion-exchange resins used at low resin/soil ratios, this relationship cannot be assumed (Barrow & Shaw, 1977; Pierzynski, 1991) and is not necessary for the assessment of bioavailable P.

Anion-Exchange Resins

The use of anion-exchange resins is the most common P-sink approach for assessing available inorganic P in soils. The procedure typically involves the use of Cl saturated resin at a 1:1 resin-to-soil ratio in 10 to 100 mL of water or weak electrolyte for 16 to 24 h (Amer et al., 1955; Olsen & Sommers, 1982). Correlations between plant response and resin extractable P are comparable or superior to correlations with chemical extraction methods (Table 7–2; Fixen & Grove, 1990).

To prevent the diffusion of P from the soil to the resin from being the rate-limiting step, resins should be intimately mixed with the soil, which creates difficulties in separating the resin from the soil for P analysis. Two approaches to the separation problem are to grind the soil to an average particle size smaller than that of the resin, which probably changes the P release characteristics of the soil, or to enclose the resin in a mesh bag, which limits resin-soil contact. Thien and Myers (1991) have used sucrose solution to facilitate the separation of soil from resin beads. The solution has a density less than the soil particles and greater than the resin beads. In a mixture of resin, soil and sucrose solution, the resin

Table 7–2. Relationship between crop uptake of P and soil test P determined by P sink and chemical extraction.

Crop	Location	Number	Soil P test[†]	Correlation coefficient	Reference
Canola	Saskatchewan	135	AEM	0.92	Qian et al., 1992
			Olsen	0.87	
Cotton	Brazil	28	AER	0.85	Raij et al., 1986
			0.02 M H_2SO_4	0.68	
Maize	Alabama	32	Fe-O Strip	0.87	Menon et al., 1989b
			AER	0.62	
			Olsen	0.81	
			Bray 1	0.74	
Rice	Brazil	8	AER	0.98	Raij et al., 1986
			AB-DTPA	0.41	
Ryegrass[‡]	New Zealand	56	AEM	0.92	Saggar et al., 1992
			Olsen	0.87	
Ryegrass	Finland	32	Fe-O Strip	0.93	Yli-Halla, 1990
			Acetic acid	0.88	
Sorghum–sudangrass	Montana	75	PST	0.91	Schaff et al., 1992
			Olsen	0.87	
Sudangrass–sorghum–barley	Colorado	23	AER	0.92	Bowman et al., 1978
			Olsen	0.88	
Wheat	China	39	Fe-O Strip	0.84	Lin et al., 1991
			AER	0.83	
			Olsen	0.83	
			Bray 1	0.56	

[†] AEM represents anion exchange membrane; AER, anion exchange resin; AB-DTPA, ammonium bicarbonate diethylene biaminepeutaacetic acid; and PST resin phytoavailability soil test.
[‡] Relationship between relative crop yield and soil P test.

beads float and can be easily separated from the soil. The sucrose has no effect on resin composition.

A recent development in the use of ion-exchange resins for assessing P release from soils is the phytoavailability soil test (PST) (Skogley et al., 1990; Yang et al., 1991). This procedure involves placing a spherical capsule containing a mixture of anion- and cation-exchange resins into a saturated soil paste for a specified period of time. The accumulation of P by the resin capsule is influenced both by the P release characteristics of the soil itself and properties controlling P diffusion through soils, unlike batch extractions where no information is obtained on diffusion through soils. The P flux is expressed in units of micromole per square centimeter per second and values are commonly 10 to 100 times greater than those reported by other workers, indicating the optimum diffusion conditions used in the procedure and the use of a mixed-bed resin rather than anion-exchange resin alone (Yang et al., 1991). The mixed-bed resin is also a closer approximation of P uptake by a root than anion-exchange resin alone. Calibration data for P with the PST is limited. Greenhouse studies indicated that the correlation between P uptake by sorghum [*Sorghum bicolor* (L.) Moench]–sudangrass [*Sorghum x drummondii* (Steudel) Millsp. & Chase] and

PST results were as good or better than those with the Olsen P soil test (Table 7–2; Schaff et al., 1992).

Ion-Exchange Resin Impregnated Membranes

A similar approach using ion-exchange resin impregnated membranes has been investigated by several researchers (Abrams & Jarrell, 1992; Qian et al., 1992: Saggar et al., 1992). Impregnation of the resin onto a plastic membrane facilitates separation of the resin beads from the soil and may eliminate the soil grinding step. Also, an extraction time as short as 15 min can be used without reducing the accuracy of predicted P availability for a wide range of soils (Qian et al., 1992). In pot studies, the resin membranes have provided a better index of P availability than conventional chemical extraction methods for canola (*Brassica napus* L.; Qian et al., 1992) and ryegrass (*Lolium perenne* L.; Saggar et al., 1992; Table 7–2).

Ion exchange membranes have the potential to estimate P availability in aquatic as well as soil environments. Edwards et al. (1993) used ion exchange membranes to obtain in situ estimates of the chemical composition of river water for two Scottish watersheds. It was suggested that direct multi-element analysis by x-ray fluorescence of ions retained on the membranes removes the need for sample storage or filtration, both of which can be sources of potential contamination and error. Thus, the membranes can provide useful information in addition to that obtained by conventional sampling (Edwards et al., 1993).

Iron-Oxide Impregnated Paper

Another P sink that has received attention is that of Fe-oxide impregnated filter paper, which has successfully estimated plant available P in a wide range of soils and management systems (Menon et al., 1989b, 1990; Sharpley, 1991). The Fe-oxide strips are prepared by exposing filter paper that has been soaked in a strong $FeCl_3$ solution to either NH_4OH solution or NH_3 vapor. Data from Myers et al. (1993) indicated that the preparation method (exposure to NH_4OH solution vs. NH_3 vapor) influenced Fe-oxide extractable P in two out of three soils tested. When a significant difference due to preparation method existed, papers exposed to NH_3 removed more P than papers exposed to NH_4OH solution. The apparent cause of this effect was unreacted $FeCl_3$ in the strips exposed to NH_3 vapor, which lowered the pH of the 0.01 M $CaCl_2$ solution used in the extraction procedure.

Myers et al. (1993) also attempted to quantify soil adhering to the Fe-oxide strips and its effect on Fe-oxide strip extractable P (Table 7–3). When there was a significant increase in soil contamination due to filter paper type, there was also a significant increase in extractable P. Subsequent work indicated that enclosing the strips in a rigid nylon mesh reduced contamination by soil.

Similarly, Perrot and Wise (1993) found that soil particles adhering to the filter paper surface, even after careful rinsing, contributed up to 85% of the Fe-oxide strip P value for P-retentive soils of high inorganic P content (690–3290 mg kg^{-1}) from past P fertilizer applications.

Table 7–3. Effects of filter paper type on soil contamination and associated extractable P in the Fe-oxide strip P determination (data from Myers et al., 1993).[†]

Paper type[‡]	Kahola soil		Tully soil		Haynie soil	
	Soil contamination	Fe-oxide P	Soil contamination	Fe-oxide P	Soil contamination	Fe-oxide P
	mg	mg kg^{-1}	mg	mg kg^{-1}	mg	mg kg^{-1}
50	4.8b	25.6b	3.9b	44.5b	1.8b	20.5b
589rr	5.5b	26.4ab	6.4b	39.4b	3.1b	20.1b
541	11.0a	30.6a	16.4a	53.2a	6.0a	29.2a

[†] Means within a column followed by the same letter are not significantly different ($P = 0.05$).
[‡] Paper types 50 and 541 are from Whatman International Ltd., Mainstone, England, and 589rr is red ribbon from Schleicher and Schuell, Keene, NH.

Adoption in Soil Testing

Wide-spread adoption of the P-sink approach for routine soil testing has not yet occurred in the USA, although parts of Brazil have used the method for the last decade (Raij et al., 1986). The Fe-oxide strip method has limited advantages compared with currently used chemical extractants. The short equilibration time, however, and multi-element assessment capabilities make the resin membranes more suitable to routine soil testing. Also, all the P-sink approaches can be used in the field to assess in situ P availability of soil or river sediment P (Edwards et al., 1993; Menon et al., 1990). Perhaps the most important advantage is that P-sink approaches operate with limited chemical extraction and are therefore more suited to a wide range of soils, irrespective of management history (Qian et al., 1992; Sharpley, 1991; Somasiri & Edwards, 1992; Yang et al., 1991). Where fertilizer history is unknown and frequent changes in fertilizer type, including rock phosphate, may have been made, it is difficult to chose the appropriate soil test. For example, Olsen P can underestimate and Bray 1 P overestimate P availability in soils amended with rock P, while P-sink approaches have provided accurate estimates (Menon et al., 1989a; Raij et al., 1986; Saggar et al., 1992). Even so, detailed field calibration and improvement in standardized methodology will be essential before any of the P-sink approaches can be used routinely to make reliable fertilizer recommendations.

ENVIRONMENTAL TESTS

For many years, soil P tests have provided farmers an indication of the economic return on a fertilizer investment and a means to evaluate the efficiency of their nutrient management plans. A large percentage of soils in areas with intensive cropping and livestock systems, however, have been found to have high or excessive soil test P, in terms of crop requirements (Fig. 7–2). Although not it's responsibility, soil testing has not been completely successful at avoiding the build-up of soil test P to levels that may be of concern to the accelerated eutrophication of surface waters. Accelerated eutrophication is the increased biological productivity of water, which can limit its use for drinking, recreation, and industry. In addition, toxins produced by some blue-green algae can pose a health risk

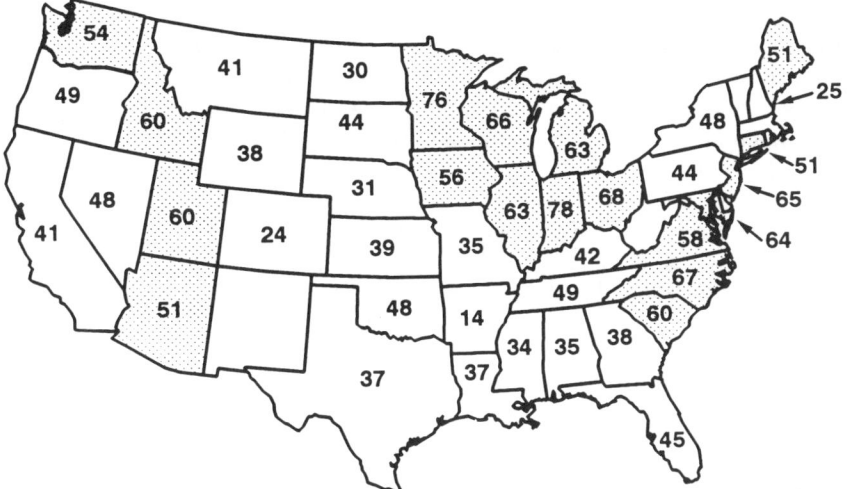

Fig. 7–2. Percentage of soil samples testing high or above for P in 1989. Highlighted states have 50% or greater of soil samples testing in the high or above range (data adapted from Potash & Phosphate Institute, 1990; Sims, 1993).

to humans and animals if consumed (Kotak et al., 1993). These toxins also contribute to taste and odor problems and interfere with the treatment of drinking water, via trihalomethane formation during water chlorination (Palmstrom et al., 1988).

The transport of bioavailable P in dissolved (DP) and particulate forms (BPP) is generally considered to promote freshwater eutrophication (Sharpley et al., 1994). While DP is mostly immediately available for algal uptake, BPP provides a variable, but long-term source of P (Peters, 1981; Sharpley et al., 1992). Thus, environmental tests for P must answer several important questions concerning excessive P levels in soil, vulnerability to release DP and transport of BPP in runoff, and their bioavailability in aquatic ecosystems. Do routine soil test extractants, designed to assess plant availability of P, measure bioavailable soil P forms important to eutrophication? If not, are other types of soil tests for P available and more appropriate? Can the long-term capacity of a soil profile to retain P from release to leaching and runoff water be determined? In a broader sense, can soil testing laboratories play a more comprehensive and proactive role in providing additional analytical or interpretive information on site vulnerability to P loss?

Phosphorus Bioavailability

The amount of P in soil, sediment, and runoff that is potentially available for algal uptake (bioavailable P) can be quantified by algal assays, which require up to 100-d incubations (Miller et al., 1978). Thus, more rapid chemical extractions, such as NaOH (Butkus et al., 1988; Dorich et al., 1980) NH_4F (Porcella et al., 1970), ion exchange resin (Huettl et al., 1979), and citrate–dithionite–bicarbonate (Logan et al., 1979), have been used to estimate bioavailable P on a routine basis.

The weaker extractants (NH_4F and NaOH) and short-term resin extractions may represent P that could be used by algae in the photic zone of lakes under aerobic conditions. In contrast, the more severe extractants (citrate–dithionite–bicarbonate) represent P that may become bioavailable under the reducing conditions found in the anoxic hypolimnion of stratified lakes.

Sharpley et al. (1991a) showed that using a wide solution/soil ratio (500:1), 0.1 M NaOH extractable P (NaOH-P) was closely related to the growth of several algal species (Fig. 7–3). The complexity of algal assay and chemical extraction methods, however, often limits their use by soil testing laboratories. For example, long assay incubation (7–100 d) and chemical extraction times (>16 h), as well as large solution volumes (>500 mL) are particularly inconvenient. As the amount of P extracted depends on ionic strength, cationic species, pH, and volume of the extractant used (Hope & Syers, 1976; Sharpley et al., 1981), these limitations will be difficult to overcome. Questions have also been raised as to the validity of relating the form or availability of P extracted by chemical solutions to P bioavailability in the aquatic environment.

Thus, Sharpley (1993) proposed a simpler method to measure bioavailable P in soil, sediment, or runoff water using Fe-oxide impregnated filter paper

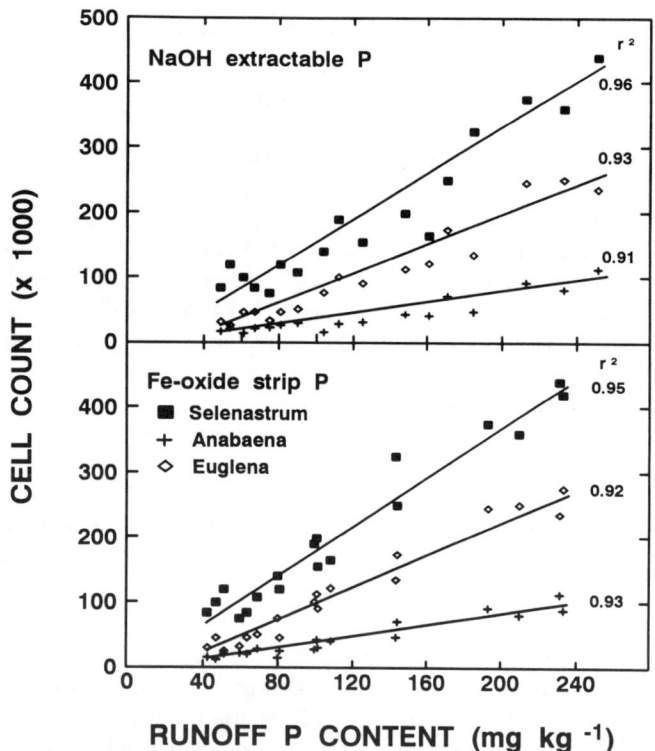

Fig. 7–3. Relationship between NaOH extractable and Fe-oxide strip P content of runoff sediment and growth of P-starved algae during a 26-d incubation.

(Fe-oxide strip) as a P-sink followed by removal of P from the Fe-oxide strip by dilute acid. Sharpley (1993) observed that the Fe-oxide strip P content of runoff was closely related to the growth of several algal species incubated for 29-d with runoff as the sole source of P (Fig. 7–3). As the strips act as a P-sink, they simulate P removal from soil or sediment-water samples by plant roots and algae. Thus, the Fe-oxide strip method has a stronger theoretical justification for its use compared with chemical extractants to estimate bioavailable P. The method may have potential use as an environmental soil P test to identify soils liable to enrich runoff with sufficient P to accelerate eutrophication.

The widespread adoption of extraction methods to estimate bioavailable P will be limited by the added laboratory and analytical complexities involved along with a lack of rigorous field testing to determine critical P levels associated with eutrophication. Routine soil tests are related to NaOH and Fe-oxide estimates of bioavailable P for soils of similar physical and chemical properties (Lin et al., 1991; Saarela, 1992; Wolf et al., 1985; Yli-Halla, 1990). For 201 soils covering a wide range of properties, however, Sharpley (1991) found 0.1 M NaOH extractable P was not closely related ($r^2 = 0.12$–0.48) to the recommended soil test P method (Fig. 7–4). When grouped in common soil orders, NaOH-P was linearly

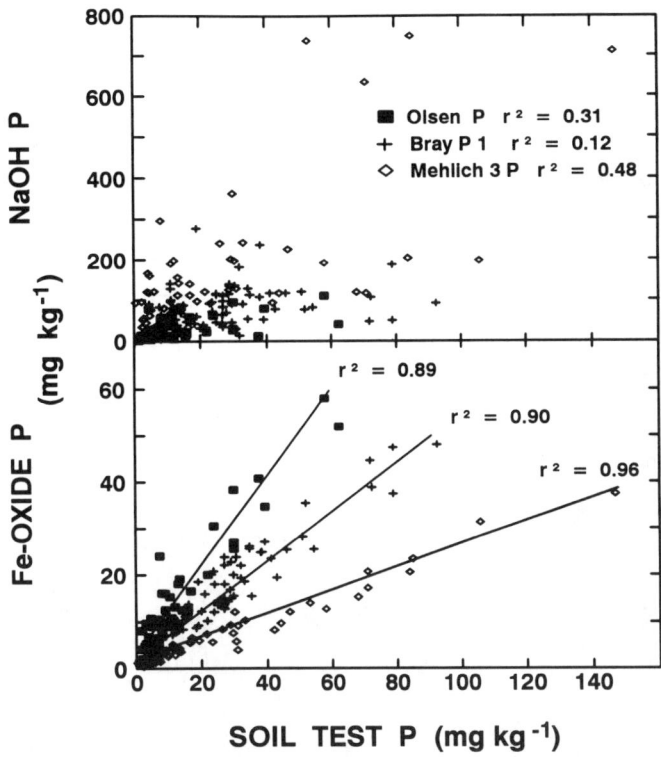

Fig. 7–4. Relationship between NaOH extractable and Fe-oxide strip P and soil test P content of 201 soils.

related to Olsen P for Entisols, Mollisols, Oxisols, and Spodosols; to Bray 1 P for Aridisols, Entisols, Inceptisols, and Mollisols; and to Mehlich 3 P for Mollisols, Spodosols, and Ultisols (Sharpley & Smith, 1993). Wolf et al. (1985) found that NaOH-P was related ($r^2 = 0.67$–0.98) to resin extractable, Olsen, Bray 1, and Mehlich 1 P content of 91 soils from the eastern USA, when grouped according to extractable Fe and Al content. Because of the ease of NaOH-P determination (17–h extraction with 0.1 M NaOH), its measurement need not be substituted by soil test P to evaluate the relative effects of agricultural management on potential bioavailable P loss in runoff.

In contrast to NaOH-P, Fe-oxide strip P was closely related to soil test P as determined by the recommended method for a given soil (Sharpley, 1991; Fig. 7–4). The differing relationship between soil test P and bioavailable P as determined by NaOH or Fe-oxide strips, results from the fact that NaOH extracts mainly amorphous and some crystalline Al and Fe phosphates, in addition to the physically sorbed P removed by the Fe-oxide strip. As the Fe-oxide strips act as a sink for P with limited chemical extraction of P, their effectiveness in estimating bioavailable P is mostly independent of soil type.

Phosphorus Sorption

From an environmental perspective, tests other than for bioavailable P will be needed. For example, with continual application of manure to agricultural soils and in development of waste water irrigation systems, estimates of the long-term capacity of a soil profile to retain P against leaching will be needed. This capacity is commonly estimated by sorption isotherms that can be used to derive sorption maxima for soil horizons (Fig. 7–5). Equilibrium P concentration (EPC_0) can also be calculated from the isotherm (Fig. 7–5). The EPC_0 is

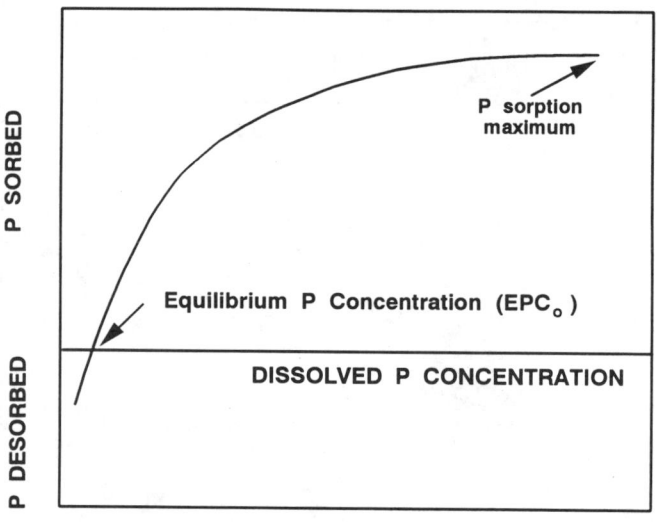

Fig. 7–5. Diagrammatic representation of P sorption isotherm showing sorption–desorption characteristics determined.

defined as the dissolved P concentration supported by a soil sample at which no net sorption or desorption occurs. Soils with high EPC_0 values have a greater tendency to desorb DP into runoff waters. If the DP concentration of runoff is greater than the EPC_0, P may be sorbed by the soil or sediment material contacted.

These sorption isotherms require equilibration of soil with a series of P solutions of increasing P concentration, normally for 24 h, and are not well adapted to routine soil testing laboratories. Bache and Williams (1971), however, suggested that a single-point isotherm could be used to estimate the P sorption maxima of soils with reasonable accuracy. This was recently confirmed by Mozaffari and Sims (1994) for surface and subsoil horizons of four Atlantic Coastal Plain soils (Fig. 7–6).

Wolf et al. (1985) reported EPC_0 was related to soil test P determined by Bray 1, Mehlich I, and Olsen methods, when 91 noncalcareous soils were grouped into coarse and medium textured soils. Similar relationships were obtained by Sharpley et al. (1982) for 66 calcareous and noncalcareous soils from across the USA. For these soils, the highest correlation ($r^2 = 0.92$) was obtained between EPC_0 and water extractable soil P (water/soil ratio of 20:1 for 5 min shaking) (Fig. 7–7).

Although the value of the EPC_0 for a given soil will be affected by solution/soil ratio and time of shaking, the relative magnitude can be predicted from a simple and rapid water extractable P test. This would eliminate the construction of P sorption isotherms in order to determine EPC_0. Use of the correlation between EPC_0 and water extractable P content, however, should be limited to a common group of soils. Furthermore, the relationship should be calibrated for a certain locality and not applied to another area without the knowledge of P-fertilization history, recalibration, and testing.

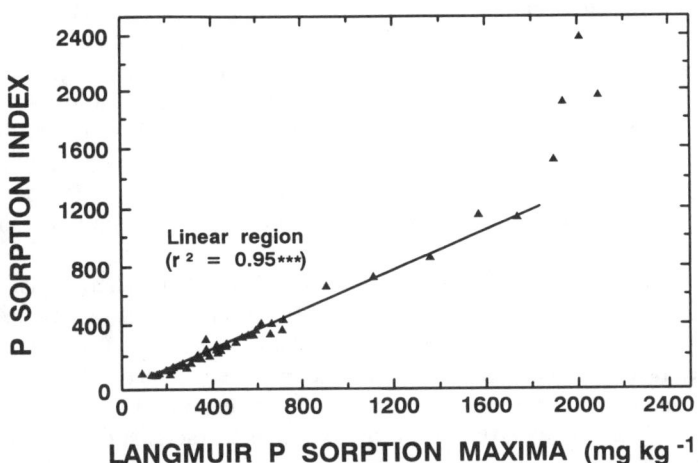

Fig. 7–6. Relationship between a P sorption index (PSI) and the P sorption maxima estimated from the Langmuir equation. Data from samples collected at seven depths in agricultural fields and field border areas of four soils (adapted from Mozaffari & Sims, 1994).

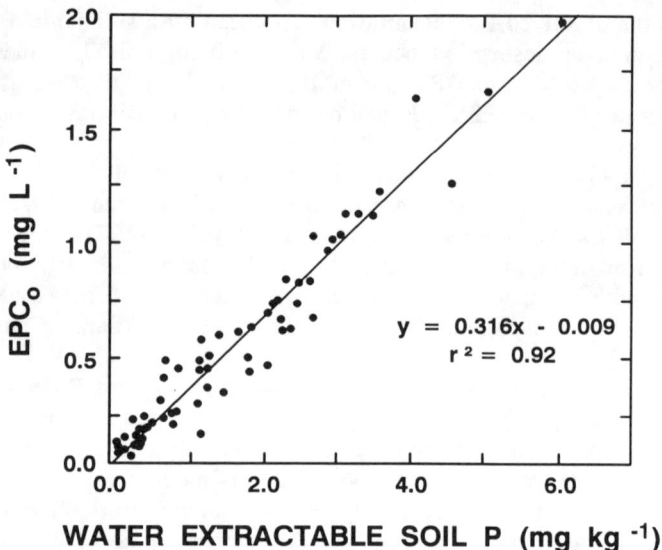

Fig. 7–7. Relationship between water extractable soil P (solution/soil ratio of 20:1 for 5 min) and equilibrium P concentration (EPC_0).

Even though there are limitations of soil type, the use of Fe-oxide strip P, P sorption index, and water extractable P offer considerable promise as surrogates for extensive soil P testing procedures that are not well adapted to a soil testing laboratory.

Phosphorus Indexing

Soil testing alone cannot assess the potential for soil P from an individual site or watershed to play a significant role in surface water eutrophication. Any environmental soil P test must be linked to site assessment of drainage, runoff, and erosion potential and management factors affecting the vulnerability for P transport from a site. Strategies to minimize P loss in runoff will be most effective if sensitive or source areas within a watershed are identified, rather than widespread implementation of general strategies over a broad area. Thus, a P indexing system was developed to identify soils vulnerable to P loss in runoff (Lemunyon & Gilbert, 1993).

The index is outlined in Tables 7–4 and 7–5. Each site characteristic affecting P loss is assigned a weighting, assuming that certain characteristics have a relatively greater effect on potential P loss than others. The P loss potential is given a value (Table 7–3), although each user must establish a range of values for different geographic areas. An assessment of site vulnerability to P loss in runoff is made by selecting the rating value for each site characteristic from the P index (Table 7–4). Each rating is multiplied by the appropriate weighting factor. Weighted values of all site characteristics are summed and site vulnerability obtained from Table 7–5.

Table 7-4. The phosphorus indexing system to rate the potential P loss runoff from site characteristics.†

Site characteristic (Weight)	Phosphorus loss potential (value)				
	None (0)	Low (1)	Medium (2)	High (4)	Very High (8)
Transport factors					
Soil erosion (1.5)¶	Negligible	<10	10–20	20–30	>30
Runoff class (0.5)	Negligible	Very low or low	Medium	High	Very High
Phosphorus source factors					
Soil P test (1.0)	Very low	Low	Medium	High	Very high
P fertilizer application rate (0.75)§	None	1–15	16–45	46–75	>76
P fertilizer application method (0.5)	None	Placed deeper than 5 cm	Incorporated immediately before crop	Incorporated >3 mo before crop or surface applied <3 mo before crop	Surface applied >3 mo before crop
Organic P source application rate (0.5)§	None	1–15	16–30	30–45	>45
Organic P source application method (1.0)	None	Injected deeper than 5 cm	Incorporated immediately before crop	Incorporated >3 mo before crop or surface applied <3 mo before crop	Surface applied >3 mo before crop

†The P indexing system was developed by the following team of scientists; J. Lemunyon, D. Goss, G. Gilbert, J. Kimble, T. Sobecki, USDA-SCS; A. Sharpley, USDA-ARS; T. Daniel, Univ. Arkansas; T. Logan, Ohio State Univ.; G. Pierzynski, Kansas State Univ.; T. Sims, Univ. Delaware; and R. Stevens, Washington State Univ.
¶Units for soil erosion are Mg ha^{-1}.
§Units for P application are kg P ha^{-1}.

A hypothetical site is used as an example, where soil erosion is 25 Mg ha^{-1} (weighting × value; 1.5 × 4 = 6), runoff class is high (0.5 × 4 = 2), soil test P is high (1 × 4 = 4), 40 kg P ha^{-1} of fertilizer (0.75 × 2 = 1.5) and 60 kg P ha^{-1} as animal manure (0.5 × 8 = 4) are broadcast in early spring prior to planting (1 × 4 = 4; 0.5 × 4 = 2). The sum of these weighted values (6, 2, 4, 1.5, 4, 2, and 4) is 23.5, which has a high site vulnerability (Table 7–5). In this hypothetical situation, conservation measures to minimize erosion and runoff as well as a P management plan should be implemented to reduce the risk of P movement and probable water quality degradation, if a P-sensitive water body is nearby.

The index is intended for use as a tool for field personnel to easily identify agricultural areas or practices that have the greatest potential to accelerate

Table 7–5. Site vulnerablity to P loss as a function of total weighted rating values from the index matrix.

Site vulnerablility	Total index rating value
Low	<10
Medium	10–18
High	19–36
Very high	>36

eutrophication. It is intended that the index will identify management options available to land users that will allow them flexibility in developing control strategies.

Implications to Soil Test Laboratories

In areas where nonpoint source pollution by P in agricultural runoff is important in eutrophication of surface waters, soil-testing laboratories could use routine soil tests to provide preliminary rankings of the bioavailable P level of soil (or sediment) and use Fe-oxide strips, P sorption index, and EPC_0 surrogates, as a special test for more intensive management programs.

Many soil testing laboratories have recommended for years that no P fertilizers be applied to soils testing very high in P, perhaps with the exception of a small amount in a starter fertilizer. The economic implications of STP-based land application programs for animal-based agricultural and municipalities with sludge disposal responsibilities, however, are highly significant. Several states (e.g., Arkansas, Delaware, Michigan, Mississippi, Ohio, Oklahoma, Pennsylvania, and Wisconsin) have attempted to identify a soil test level where no P from manures or fertilizers is recommended beyond that which would be removed from the field in the harvested portion of the crop (Table 7–6). In many situations, this would require that no manure or sewage sludge be applied and that alternative end-uses be developed. This is a clear example of the need for an integrated approach to the use of soil test P information and the expanding role that soil test laboratories will play in this process.

Advances in the interpretation of soil test P results for environmental problems will require continued innovation. It is essential that the long-term, proactive role of soil testing laboratories in the development of sampling procedures, analytical methods, and practical recommendations for efficient P management for crop production be applied to environmental management of soil P. Soil testing programs have a unique opportunity to coordinate the efforts of many of the participants needed to develop effective, environmentally sound P management programs.

In many states, the traditional interactions of soil testing laboratories with university-based research and extension soil scientists and agronomists have expanded into closer working relationships with advisory agencies (U.S. Soil Conservation Service and local conservation districts), crop and environmental

Table 7–6. Soil P interpretations and management guidelines currently in use.

State	Critical value	Management recommendation	Rationale†
Arkansas	150 mg kg^{-1} Mehlich 3 P	At or above 150 mg kg^{-1} STP: 1. Apply no P from any source. 2. Provide buffers next to streams. 3. Overseed pastures with legumes to aid P removal. 4. Provide contstant soil cover to minimize erosion.	CV: data from Ohio with sewage sludge. MR: reduce P levels and minimize movement of P from field.
Delaware	120 mg kg^{-1} Mehlich 1 P	Above 120 mg kg^{-1} STP: Apply no P from any source until STP is significantly reduced.	CV: greater P loss potential from high P soils. MR: protect water quality by minimizing further P accumulations
Ohio	150 mg kg^{-1} Bray 1 P	Above 150 mg kg^{-1} STP: 1. Institute practices to reduce erosion. 2. Reduce or eliminate P additions.	CV: greater P loss potential from high P soils as well as role of high soil P in zinc deficiency. MR: protect water quality by minimizing further P accumulations.
Oklahoma	130 mg kg^{-1} Mehlich 3 P	30 to 130 mg kg^{-1} STP: Half P rate on >8% slopes. 130 to 200 mg kg^{-1} STP: Half P rate on all soils and institute practices to reduce runoff and erosion. Above 200 mg kg^{-1} STP: P rate not to exceed crop removal.	CV: greater P loss potential from high P soils. MR: protect water quality, minimize further soil P accumulation, and maintain economic viability.
Michigan	75 mg kg^{-1} Bray 1 P	Above 75 mg kg^{-1} STP: P application must not exceed crop removal. Above 150 mg kg^{-1} STP: Apply no P from any source.	CV: minimize P loss by erosion or leaching in sandy soils. MR: protect water quality and encourage wider distribution of manures.
Wisconsin	75 mg kg Bray 1 P	Above 75 mg kg STP: 1. Rotate to P demanding crops. 2. Reduce manure application rates. Above 150 mg kg STP: Discontinue manure applications.	CV: at that level, soils will remain non-responsive to applied P for 2–3 years. MR: Minimize further P accumulations.

† CV represents critical value rationale and MR, management recommedation rationale.

consulting firms, and state and federal regulatory organizations. Soil testing programs, through sustained and creative efforts, can contribute greatly to the development of conceptually sound and technically feasible solutions to the complex problems of nonpoint source pollution by soil P.

REGIONAL SOIL TESTING

Currently, most soil testing recommendations made by state or university laboratories are confined to individual states; private laboratories often operate over an entire region, or even on a national basis. Conceptually, it makes little sense to approach soil testing from a political perspective (state borders) instead of a physiographic one; indeed the current structure often inhibits innovation in soil testing. Soils, cropping systems, and agricultural business (e.g., fertilizer sales and application, animal-based agriculture) certainly change independently of political borders and it would be much more appropriate if organized efforts at soil testing and interpretation did so as well. This would allow for better tailoring of analytical procedures to appropriate soils and for a more focused, coherent strategy in the development and modification of soil-testing based nutrient management programs.

University and state soil testing laboratories, however, are often constrained by administrative and budgetary structures that allow little flexibility in expanding (or contracting) the geographic areas that they must provide services to. Private soil testing laboratories often face a different problem. Many of the larger private laboratories operate at a national level or over extensive geographic regions (e.g., the Midwest). Economics, however, have usually dictated that a private laboratory must use a fairly standardized set of protocols to serve many states with dissimilar soils and cropping systems. Simply put, private laboratories that serve multiple states usually cannot afford to offer an unlimited number of physiographically appropriate soil testing methods. This has resulted in an increased interest on their part in universal soil testing extractants that are effective for multiple elements across a wide range of soils. Jones (1990) summarized the current status of universal extractants and listed the most common as AB-DTPA, Mehlich 1, and Mehlich 3.

Individuals responsible for soil testing have, of course, long recognized the value of regional coordination of their efforts. To date, the most common response to this need has been the formation of regional soil testing committees that meet on an annual basis to discuss common problems and devise unified methodological and interpretive strategies. Some soil testing laboratories participate in more than one regional committee, enhancing somewhat the interchange of ideas on a national basis. Examples of these committees and the participating laboratories include the Northeast Coordinating Committee on Soil Testing (NEC-67: Connecticut, Delaware, Maine, Maryland, Massachusetts, New Hampshire, New Jersey, New York, Pennsylvania, Rhode Island, Vermont, and West Virginia), the mid-Atlantic Soil Testing and Plant Analysis Workgroup (MASTPAWG) that has participants from both the private (A&L Laboratories, Brookside Laboratory, and AGRICO) and state-university sectors (Delaware,

Georgia, Maryland, New Jersey, South Carolina, Pennsylvania, Virginia, and West Virginia); the North Central Soil Testing Committee (NCR-13: Illinois, Indiana, Iowa, Kansas, Michigan, Minnesota, Missouri, Nebraska, North Dakota, Ohio, Pennsylvania, South Dakota, and Wisconsin), and the Southern Extension and Research Activities (SERA-6: Alabama, Florida, Georgia, Kentucky, Louisiana, North Carolina, Oklahoma, South Carolina, and Tennessee) (Table 7–1). All of these committees are rather informal in nature and are perhaps best viewed as technical discussion groups that promote gradual, voluntary change in soil testing practices. Most conduct sample exchanges to verify analytical accuracy and identify differences in interpretation due to factors such as soils, climate, cropping systems, or philosophy of the individual state. This is an important point to note—any innovations in soil testing that may arise based on the efforts of the regional committee must still undergo rather critical review at the state level by research and extension personnel responsible for nutrient recommendations before implementation. Historically, as mentioned above, this has meant that changes in soil testing methods or interpretation will be slow, and only occur when they have been well-documented to be superior to current practices.

Innovations in soil testing for P, for either agronomic or environmental purposes, will, for the most part, be enhanced by the activities of regional soil testing committees. Despite the generally slow nature of changes in soil testing practices, most individuals involved in these committees are quite interested in any advances in soil testing that will improve agronomic profitability and minimize any negative environmental effects from nutrient use. Further, they are often in close contact with the soil scientists conducting basic or applied research on the chemistry of soil P, the transport of P in erosion, runoff, or drainage, and in the formulation of best management practices for P in agriculture. As mentioned earlier, many also interact at the local level with representatives of agribusiness, extension (USDA-SCS and Cooperative Extension), and regulatory agencies (state and U.S. Environmental Protection Agency), and with community action groups. Regional committees provide the opportunity to review these local interactions and perhaps translate them into broader scale research or educational efforts. As an example, NEC-67 recently published a regional bulletin, *Recommended Soil Testing Procedures for the Northeastern United States* (Sims & Wolf, 1991). This bulletin reviewed the basis for testing methods and provided specific procedures to follow when conducting the tests. Because several members of NEC-67 were also involved in NCR-13 and the MASTPAWG, the methodology for tests common to those regions was incorporated promoting standardization of analytical procedures across regions. More importantly, NEC-67 is now reviewing issues such as differences among northeastern states in fertilizer and manure recommendations, approaches to manage high P soils, testing procedures for heavy metals in soils, and the regulatory aspects of soil testing for elements that have possible environmental effects. Expanded regional efforts such as these, that focus on interpretation and resolution of philosophical differences on soil P testing, as well as analytical methods, are now a major need in all regions of the USA.

CONCLUSIONS

We have attempted to identify and answer questions concerning the development of innovative soil extraction, extrapolation, and interpretation approaches to recommend P applications that are agronomically, economically, and environmentally sound. As questions about the role of agriculture in P enrichment and eutrophication of surface waters are being asked more frequently, soil testing will be required to assess the availability of soil P to loss in runoff and drainage water as well as to crop uptake. For both agronomic and environmental soil testing for P, collection of a representative sample will be essential to obtaining reliable recommendations. Although sampling strategies will be similar, the depth of soil sampling for agronomic (0–20 cm) and environmental (0–5 cm) recommendations will reflect the respective depths relevant to crop uptake and release to runoff.

Several questions regarding the adoption of P-sink approaches in soil testing for P still remain unresolved. Phosphorus-sink approaches simulate P uptake more closely than chemical extraction of P and should, thus, provide a close representation of soil P availability. In addition, their adaptation to short-time equilibration (<1 h), multi-element analysis (anions and cations), and to a wide range of soil types and management systems, suggest they have potential as a universal soil test. Widespread adoption of P-sink approaches in both agronomic and environmental soil testing, however, is limited by a lack of field calibration and interpretation data.

Calibration research that reflects changes in tillage practices, crop hybrids, and fertilizer sources or application techniques, is needed to develop reliable and innovative agronomic soil testing for P. Due to inadequate funding of long-term field calibration research, answers will probably be provided by greenhouse studies under artificial conditions. These questions are also relevant to environmental soil testing for P, where spatial and temporal variations in climate and soil properties result in a highly variable response of surface waters to specific inputs of P. Questions raised by the lack of field calibration or interpretation data relating soil levels to the trophic state or response of associated P-sensitive surface waters, mean that reliable recommendations on the environmental impacts of agricultural P management from soil testing, are still some time away. Until these questions can be answered, surrogate measurement may provide useful indicators of a soil's eutrophication potential (water extractable P or EPC_0) or P loading capacity (P sorption index or clay content).

Finally, innovations in soil testing for P should be developed across regional rather than within state boundaries. This will more clearly focus local analytical, interpretational, and philosophical differences toward coherent soil test strategies for P. As soil testing in general and farm advisory services in particular, gain additional responsibility to develop farm nutrient management plans that are agronomically, economically, and environmentally sound, innovative methods will be essential.

REFERENCES

Abrams, M.M., and W.M. Jarrell. 1992. Bioavailability index for phosphorus using nonexchange resin impregnated membranes. Soil Sci. Soc. Am. J. 56:1532–1537.

Adepoju, A.Y., P.F. Pratt, and S.V. Mattigood. 1982. Availability and extractability of phosphorus from soils having high residual phosphorus. Soil Sci. Soc. Am. J. 46:583–588.

Amer, F., D.R. Bouldin, C.A. Black, and F.R. Duke. 1955. Characterization of soil phosphorus by anion exchange resin and adsorption by P-32 equilibration. Plant Soil 6:391–408.

Aquino, B.F., and R.G. Hanson. 1984. Soil phosphorus supplying capacity evaluated by plant removal and available phosphorus extraction. Soil Sci. Soc. Am. J. 48:1091–1096.

Bache, B.W., and E.G. Williams. 1971. A phosphate sorption index for soils. J. Soil Sci. 22:289–301.

Barrow, N.J., and T.C. Shaw. 1977. Factors affecting the amount of phosphate extracted from soil by anion exchange resin. Geoderma 18:309–323.

Bates, T.E. 1990. Prediction of phosphorus availability from 88 Ontario soils using five phosphorus soil tests. Commun. Soil Sci. Plant Anal. 21:1009–1024.

Beegle, D.B., and T.C. Oravec. 1990. Comparison of field calibrations for Mehlich 3 P and K with Bray-Kurtz P1 and ammonium acetate K for corn. Commun. Soil Sci. Plant Anal. 21:1025–1036.

Bowman, R.A., S.R. Olsen, and F.S. Watanabe. 1978. Greenhouse evaluation of residual phosphate by four phosphorus methods in neutral and calcareous soils. Soil Sci. Soc. Am. J. 42:451–454.

Bray, R.H., and L.T. Kurtz. 1945. Determination of total, organic, and available forms of phosphorus in soils. Soil Sci. 59:39–45.

Butkus, S.R., E.B. Welch, R.R. Horner, and D.E. Spyridakis. 1988. Lake response modeling using biologically available phosphorus. J. Water Pollut. Contol Fed. 60:1663–1669.

Combs, S.M., and S.W. Burlington. 1992. Results of the 1986–1990 Wisconsin soil test summary. Univ. of Wisconsin Memo. Rep. Dep. Soil Sci. Univ. of Wisconsin, Madison, WI.

Dorich, R.A., D.W. Nelson, and L.E. Sommers. 1980. Algal availability of sediment phosphorus in drainage water of the Black Creek watershed. J. Environ. Qual. 9:557–563.

Edwards, T., B. Ferrier, and R. Harriman. 1993. Preliminary investigation on the use of ion-exchange resins for monitoring river water composition. Sci. Total Environ. 135:27–36.

Ellis, B.G., and R.A. Olson. 1986. Economic, agronomic and environmental implications of fertilizer recommendations. North Central Regional Res. Publ. 310. Michigan State Univ. Agric. Exp. Stn., East Lansing, MI.

Fixen, P.E., and J.H. Grove. 1990. Testing soils for phosphorus. p. 141–180. In R.L. Westerman (ed.) Soil testing and plant analysis. 3rd ed. SSSA Book Ser. 3. SSSA, Madison, WI.

Gilbertson, C.B., F.A. Norstadt, A.C. Mathers, R.F. Holt, A.P. Barnett, T.M. McCalla, C.A. Onstad, and R.A. Young. 1979. Animal waste utilization on cropland and pastureland. USDA Utilization Res. Rep. 6. U.S. Gov. Print. Office, Washington, DC.

Griffith, D.R., J.V. Mannering, and W.C. Moldenhauer. 1977. Conservation tillage in the eastern Corn Belt. J. Soil Water Conserv. 32:20–28.

Guertal, E.A., D.J. Eckert, S.J. Traina, and T.J. Logan. 1991. Differential phosphorus retention in soil profiles under no-till crop production. Soil Sci. Soc. Am. J. 55:410–413.

Hanlon, E.A. and G.V. Johnson. 1984. Bray/Kurtz, Mehlich III, AB/DTPA, and ammonium acetate extractions of P, K, and Mg in four Oklahoma soils. Commun. Soil Sci. Plant Anal. 15:277–294.

Hooker, M.L. 1976. Soil sampling intensities required to estimate available N and P in five Nebraska soil types. M.S. thesis. Univ. Nebraska, Lincoln (Catalog no. LD 3656 H665X 1976).

Hope, G.D., and J.K. Syers. 1976. Effects of solution to soil ratio on phosphate sorption by soils. J. Soil Sci. 27:301–306.

Huettl, P.J., R.C. Wendt, and R.B. Corey. 1979. Prediction of algal available phosphorus in runoff suspension. J. Environ. Qual. 4:541–548.

James, D.W., and K.L. Wells. 1990. Soil sample collection and handling:Technique based on source and degree of field variability. p. 25–44. In R.L. Westerman (ed.) Soil testing and plant analysis. 3rd ed. SSSA Book Ser. 3. SSSA, Madison, WI.

Jones, J.B. 1990. Universal soil test extractants: Their composition and use. Commun. Soil Sci. Plant Anal. 21:1091–1102.

Jones, C.A., C.V. Cole, A.N. Sharpley, and J.R. Williams. 1984a. A simplified soil and plant phosphorus model: I. Documentation. Soil Sci. Soc. Am. J. 48:800–805.

Jones, C.A., A.N. Sharpley, and J.R. Williams. 1984b. A simplified soil and plant phosphorus model: III. Testing. Soil Sci. Soc. Am. J. 48:810–813.

Kamprath, E.J., and M.E. Watson. 1980. Conventional soil and tissue tests for assessing the phosphorus status of soils. p. 433–469. In F.E. Khawsaneh et al. (ed.) The role of phosphorus in agriculture. ASA, CSSA, and SSSA, Madison, WI.

Kitchen, N.R., J.L. Havlin, and D.G. Westfall. 1990. Soil sampling under no-till banded phosphorus. Soil Sci. Soc. Am. J. 54:1661–1665.

Kotak, B.G., S.L. Kenefick, D.L. Fritz, C.G. Rousseaux, E.E. Prepas, and S.E. Hrudey. 1993. Occurrence and toxicological evaluation of cyanobacterial toxins in Alberta lakes and farm dugouts. Water Res. 27:495–506.

Labhsetwar, V.K., and P.N. Soltanpour. 1985. A comparison of NH_4HCO_3–DTPA, $CaCl_2$, and Na_2–EDTA soil tests for phosphorus. Soil Sci. Soc. Am. J. 49:1437–1440.

Lemunyon, J.L., and R.G. Gilbert. 1993. Concept and need for a phosphorus assessment tool. J. Prod. Agric. 6:483–486.

Lin, T.H., S.B. Ho, and K.H. Houng. 1991. The use of iron oxide-impregnated filter paper for the extraction of available phosphorus from Taiwan soils. Plant Soil 133:219–226.

Lins, I.D, F.R. Cox, and J.J. Nicholaides. 1985. Optimizing phosphorus fertilization rates for soybeans grown on Oxisols and associated Entisols. Soil Sci. Soc. Am. J. 49:1457–1460.

Logan, T.J., T.O. Oloya, and S.M. Yaksich. 1979. Phosphate characteristics and bioavailability of suspended sediments from streams draining into Lake Erie. J. Great Lakes. Res. 5:112–123.

McCallister, D.I., C.A. Shapiro, W.R. Raun, F.N. Anderson, G.W. Rehm, O.P. Englestead, M.P. Russelle, and R.A. Olson. 1987. Rate of phosphorus and potassium buildup/decline with fertilization for corn and wheat on Nebraska Mollisols. Soil Sci. Soc. Am. J. 51:1646–1652.

McCollum, R.E. 1991. Buildup and decline in soil phosphorus: 30-year trends on a Typic Umbraquult. Agron. J. 83:77–85.

McIntosh, J.L. 1969. Bray and Morgan soil test extractants modified for testing acid soils from different parent materials. Agron. J. 61:259–265.

Mehlich, A. 1953. Determination of P, Ca, Mg, K, Na and NH_4. North Carolina Soil Testing Div. Mimeo. North Carolina Soil Testing Div., Raleigh, NC.

Mehlich, A. 1984. Mehlich 3 soil test extractant: A modification of Mehlich 2 extractant. Commun. Soil Sci. Plant Anal. 15:1409–1416.

Menon, R.G., S.H. Chien, and L.L. Hammond. 1989a. Comparison of Bray 1 and P_i tests for evaluating plant-available phosphorus from soils treated with different partially acidulated phosphate rocks. Plant Soil 114:211–216.

Menon, R.G., S.H. Chien, L.L. Hammond, and B.R. Arora. 1990. Sorption of phosphorus by iron oxide-impregnated filter paper (P_i soil test) embedded in soils. Plant Soil 126:287–294.

Menon, R.G., L.L. Hammond, and H.A. Sissingh. 1989b. Determination of plant-available phosphorus by the iron hydroxide-impregnated filter paper (P_i) soil test. Soil Sci. Soc. Am. J. 52:110–115.

Miller, W.E., J.C. Greene, and T. Shiroyarna. 1978. The *Selenastrum capricornutum* Printz algal assay bottle test and data interpretation protocol. USEPA Tech. Rep. EPA-600/9-78-018. USEPA, Covallis, OR.

Morgan, M.F. 1941. Chemical soil diagnosis by the universal testing system. Connecticut Agric. Exp. Stn. Bull. 450. New Haven.

Motschall, R.M., and T.C. Daniel. 1982. A soil sampling method to identify critical manure management areas. Trans. ASAE. 25:1641–1645.

Mozaffari, M., and J.T. Sims. 1994. Phosphorus availability and sorption in an Atlantic Coastal Plain watershed dominated by animal-based agriculture. Soil Sci. 157:97–107.

Myers, R.G., G.M. Pierzynski, and S.J. Thien. 1993. Improving the FeO-Sink analysis for soil phosphorus. p. 280. *In* Agronomy abstracts. ASA, Madison, WI.

Novais, R., and E.J. Kamprath. 1978. Phosphorus supplying capacities of previously heavily fertilized soils. Soil Sci. Soc. Am. J. 42:931–935.

Olsen, S.R., C.V. Cole, F.S. Watanabe, and L.A. Dean. 1954. Estimation of available phosphorus in soils by extraction with sodium bicarbonate. USDA Circ. 939 U.S. Gov. Print. Office, Washington, DC.

Olsen, S.R., and L.E. Sommers. 1982. Phosphorus. p. 403–429. *In* A.L. Page et al. (ed.) Methods of soil analysis. Agron. Monogr. 9. Part 2. 2nd ed. ASA and SSSA, Madison, WI.

Olson, R.A., F.N. Anderson, K.D. Frank, P.H. Grabouski, G.W. Rehm, and C.A. Shapiro. 1987. Soil testing interpretations: Sufficiency vs. build-up and maintenance. p. 41–52. *In* J.R. Brown (ed.) Soil testing: Sampling, correlation, calibration, and interpretation. SSSA Spec. Publ. 21. SSSA, Madison, WI.

Palmstrom, N.S., R.E. Carlson, and G.D. Cooke. 1988. Potential links between eutrophication and formation of carcinogens in drinking water. Lake Reserv. Manage. 4:1–15.

Perrot, K.N., and R.G. Wise. 1993. An evaluation of some aspects of the iron oxide-impregnated filter paper (Pi) test for available soil phosphorus with New Zealand soils. N.Z. J. Agric. Res. 36:157–162.

Peters, R.H. 1981. Phosphorus availability in Lake Memphremagog and its tributaries. Limnol. Oceanogr. 26:1150–1161.

Pierzynski, G.M. 1991. The chemistry and mineralogy of phosphorus in excessively fertilized soils. Crit. Rev. Environ. Control 21:265–295.

Pierzynski, G.M., T.J. Logan, and S.J. Traina. 1990. Phosphorus chemistry and mineralogy in excessively fertilized soils: Solubility equilibria. Soil Sci. Soc. Am. J. 54:1589–1595.

Porcella, D.B., J.S. Kumazar, and E.J. Middlebrooks. 1970. Biological effects on sediment-water nutrient interchange. J. Sanit. Eng. Div., Proc. Am. Soc. Civ. Eng. 96:911–926.

Potash and Phosphate Institute. 1990. Soil test summaries: Phosphorus, potassium, and pH. Better Crops 74:16–19.

Qian, P., J.J. Schoenau, and W.Z. Huang. 1992. Use of ion exchange membranes in routine soil testing. Commun. Soil Sci. Plant Anal. 23:1791–1804.

Raij, B. van, J.A. Quaggio, and N.M. de Silva. 1986. Extraction of phosphorus, potassium, calcium, and magnesium from soils by an ion-exchange resin procedure. Commun. Soil Sci. Plant Anal. 17:547–566.

Saarela, I. 1992. A simple diffusion test for soil phosphorus availability. Plant Soil 147:115–126.

Saggar, S., M.J. Hedley, R.E. White, P.E.H. Gregg, K.W. Perrot, and I.S. Conforth. 1992. Development and evaluation of an improved soil test for phosphorus. 2. Comparison of the Olsen and mixed cation-anion exchange resin tests for predicting the yield of ryegrass growth in pots. Fert. Res. 33:135–144.

Schaff, B.E., E.O. Skogley, J.W. Bauder, and D.J. Sieler. 1992. Resin capsule adsorption of P for predicting plant response and P uptake. p. 290. In Agronomy abstracts. ASA, Madison, WI.

Shapiro, C.A. 1988. Soil sampling fields with a history of fertilizer bands. In Soil Science News. Vol 10. Nebraska Coop. Ext. Serv., Univ. of Nebraska, Lincoln.

Sharpley, A.N. 1991. Soil phosphorus extracted by iron-aluminum-oxide-impregnated filter paper. Soil Sci. Soc. Am. J. 55:1038–1041.

Sharpley, A.N. 1993. An innovative approach to estimate bioavailable phosphorus in agricultural runoff using iron oxide-impregnated paper. J. Environ. Qual. 22:597–601.

Sharpley, A.N., L.R. Ahuja, M. Yamamoto, and R.G. Menzel. 1981. The kinetics of phosphorus desorption from soil. Soil Sci. Soc. Am. J. 45:493–496.

Sharpley, A.N., B.J. Carter, B.J. Wagner, S.J. Smith, E.L. Cole, and G.A. Sample. 1991b. Impact of long-term swine and poultry manure applications on soil and water resources in eastern Oklahoma. Oklahoma Agric. Exp. Stn. Tech. Bull. T-169. Oklahoma State Univ., Stillwater.

Sharpley, A.N., S.C. Chapra, R. Wedepohl, J.T. Sims, T.C. Daniel, and K.R. Reddy. 1994. Managing agricultural phosphorus for protection of surface waters: Issues and options. J. Environ. Qual. 23:437–451.

Sharpley, A.N., L.W. Reed, and D.K. Simmons. 1982. Relationships between available soil phosphorus forms and their role in water quality modeling. Oklahoma Agric. Exp. Stn. Tech. Bull. T-157. Oklahoma State Univ., Stillwater.

Sharpley, A.N., and S.J. Smith. 1993. Application of phosphorus bioavailability indices to agricultural runoff and soils. p. 43–57. In K. Hoddinott and T.A. O'Shay (ed.) Application of agricultural analysis in environmental studies. ASTM STP 1162. Am. Soc. Techniques and Materials, Philadelphia, PA.

Sharpley, A.N., S.J. Smith, and R. Bain. 1993. Nitrogen and phosphorus fate from long-term poultry litter applications to Oklahoma soils. Soil Sci. Soc. Am. J. 57:1131–1137.

Sharpley, A.N., S.J. Smith, O.R. Jones, W.A. Berg, and G.A. Coleman. 1992. The transport of bioavailable phosphorus in agricultural runoff. J. Environ. Qual. 21:30–35.

Sharpley, A.N., W.W. Troeger, and S.J. Smith. 1991a. The measurement of bioavailable phosphorus in agricultural runoff. J. Environ. Qual. 20:235–238.

Sibbesen, E. 1978. An investigation of the anion-exchange resin method for soil phosphate extraction. Plant Soil 50:305–321.

Sims, J.T. 1989. Comparison of Mehlich 1 and Mehlich 3 extractants for P, K, Ca, Mg, Mn, Cu, and Zn in Atlantic Coastal Plain Soils. Commun. Soil Sci. Plant Anal. 20:1707–1726.

Sims, J.T. 1992. Environmental management of phosphorus in agriculture and municipal wastes. p. 59–64. In F.J. Sikora (ed.) Future directions for agricultural phosphorus research. Natl. Fert. Environ. Res. Ctr., TVA Muscle Shoals, AL.

Sims, J.T. 1993. Environmental soil testing for phosphorus. J. Prod. Agric. 6:501–507.

Sims, J.T., and A.M. Wolf (ed.). 1991. Recommended soil testing procedures for the northeastern United States. Delaware Agric. Exp. Stn. Bull. 493. Delaware Agric Exp. Stn., Newark.

Skogley, E.O., S.J. Georgitis, J.E. Yang, and B.E. Schaff. 1990. The phytoavailability soil test—PST. Commun. Soil Sci. Plant Anal. 21:1229–1243.

Soltanpour, P.N., and A.P. Schwab. 1977. A new soil test for simultaneous extraction of macronutrients and micronutrients in alkaline soils. Commun. Soil Sci. Plant Anal. 8:195–207.

Somasiri, L.L.W., and A.C. Edwards. 1992. An ion exchange resin method for nutrient extraction of agricultural advisory soil samples. Commun. Soil Sci. Plant Anal. 23:645–657.

Thien, S.J., and R. Myers. 1991. Separating ion-exchange resin from soil. Soil Sci. Soc. Am. J. 55:890–892.

van der Zee, S.E.A.T.M., L.G.J. Fokkink, and W.H. van Riemsdijk. 1987. A new technique for assessment of reversibly adsorbed phosphate. Soil Sci. Soc. Am. J. 51:599–604.

Ward, R., and D.F. Leikam. 1986. Soil sampling techniques for reduced tillage and band fertilizer application. *In* Great Plains Soil Fertility Workshop, Denver, CO. 4–5 Mar. 1986. Ward Laboratories, Kearney, NE.

Whitney, D.A. 1982. Soil sampling techniques under reduced tillage systems. p. 279. *In* Agronomy abstracts. ASA, Madison, WI.

Wolf, A.M., and D.E. Baker. 1985. Comparisons of soil test phosphorus by Olsen, Bray P1, Mehlich 1, and Mehlich III methods. Commun. Soil Sci. Plant Anal. 16:467–484.

Wolf, A.M., D.E. Baker, H.B. Pionke, and H.M. Kunishi. 1985. Soil tests for estimating labile, soluble, and algae-available phosphorus in agricultural soils. J. Environ. Qual. 14:341–348.

Yang, J.E., E.O. Skogley, S.J. Georgitis, B.E. Schaff, and A.H. Ferguson. 1991. Phytoavailability soil test: Development and verification of theory. Soil Sci. Soc. Am. J. 55:1358–1365.

Yli-Halla, M. 1990. Comparison of a bioassay and three chemical methods for determination of plant-available P in cultivated soils of Finland. J. Agric. Sci. Finl. 62:213–319.

8 Innovative Soil Phosphorus Availability Indices: Assessing Organic Phosphorus

H. Tiessen and J. W. B. Stewart
University of Saskatchewan
Saskatoon, Canada

A. Oberson
Centro Internacional de Agricultura Tropical
Cali, Colombia

Organic phosphorus (P_o) accounts for 20 to 60% of total P in most mineral soils. Although plants only take up inorganic phosphate (P_i), this phosphate pool is known to be replenished in part by the mineralization of P_o. The biological and biochemical reactions responsible for the mineralization of P_o are well understood in vitro and in experiments involving controlled microcosms where controlling factors could be examined independently, but in soils, where P_o undergoes a multitude of stabilizing reactions, most evidence for the mechanisms, rates and extent of P_o mineralization are indirect. No direct measurements for the availability of P_o, analogous to the soil tests for P_i, are known, and even the determination of P_o relies on an indirect method. Only P_i reacts with the standard molybdate reagent, and P_o is estimated indirectly as the difference between, molybdate-reactive P before and after an oxidative treatment of a soil extract. The direct determination of P_o by ^{31}P nuclear magnetic resonance has so far contributed little towards quantifying available P_o, and the characterization of P_o availability relies greatly on the researchers' skill to combine circumstantial evidence. In this chapter, we will discuss the short- and long-term dynamics of P_o in soils, and in the soil-plant system. Current experimental methods used for the study of P_o in soil and the associated problems are critically examined. Finally, the possibility of using laboratory or other tests as a means of predicting the availability of soil P_o to plants is explored.

ORGANIC PHOSPHORUS IN SOIL

Organic matter mineralization plays an important role in the availability of soil P. An understanding of P_o availability should therefore be based on knowledge of the factors that determine the stability of organic matter and P_o in soils.

Copyright © 1994 Soil Science Society of America, 677 S. Segoe Rd., Madison, WI 53711, USA.
Soil Testing: Prospects for Improving Nutrient Recommendations, SSSA Special Publication 40.

In the biosphere, approximately one-half of the total P is in organic form. Phosphate diesters are abundant as a structural constituent in macromolecules such as in nucleic acids and the phospholipids of biomembranes. In monoester form, P is found in a variety of compounds, which account for less than one-half of the P_o in growing bacteria, fungi, and plants (Anderson, 1980; Beever & Burns, 1980; Bieleski, 1973). One of the monoesters is inositol hexaphosphate (IHP), which functions as a storage compound for P in seeds, and occurs in minor amounts in microorganisms.

The distribution of organic phosphates in soils and soil extracts is quite different from that found in organisms, with monoesters being by far the most dominant form. The different composition of soil P_o has been explained by the differential stabilization of certain compound classes in the soil. When P_o is released into the soil environment by secretion or cell lysis, it may be recycled into the biomass immediately, or stabilized in the soil matrix. The mechanism of stabilization may be through the C moiety, or in certain cases through interactions of the P group, analogous to the reactions for P_i. The latter reaction will be most likely to occur with low-molecular weight phosphomonoesters, in which the P-moiety determines compound reactivity. The presence of several monoester groups will further enhance sorption. Thus, inositol hexaphosphate has been shown to adsorb more strongly than orthophosphate to Fe-oxide rich soil (Anderson et al., 1974). The sorption mechanism was confirmed using clays coated with short-range ordered Fe and Al hydroxides (Shang et al., 1990, 1992). This strong affinity to soil of the inositol-phosphates may explain why up to 50% of total extractable P_o can consist of various inositol phosphates, while they are much less abundant in organisms. Phosphodiesters or monoesters with larger molecular weight are less easily adsorbed in such a way that the P group is blocked from attack by phosphatase enzymes, and they are therefore less stable in the soil environment. Such differential stabilization and accumulation determines the P_o composition of most soils, which typically contain >90% as monoester P with only traces of diesters, triesters, and phosphatidyl-choline, as identified by ^{31}P nuclear magnetic resonance (Condron et al., 1985). Among the compounds, inositol phosphates of various kinds as well as several other monoesters are common, while phospholipids and nucleic acids occur in rather small amounts (Kowalenko, 1978).

Accumulation precludes cycling of the P_o compound, and only compounds that are held less strongly by the soil matrix are likely to participate in the biological cycle. Extractability of P_o from soils has, therefore, been used as an indicator of bioavailability and turnover. Bicarbonate-extractable P_o for instance has been found to be easily mineralizable, and it may contribute to plant available P during one growing season (Bowman & Cole, 1978b; O'Halloran et al., 1986; Gahoonia & Nielsen, 1992). More insoluble forms of P_o involved in the long-term transformations of P in soils are extractable with sodium hydroxide (Batsula & Krivonosova, 1973), but these may also contribute directly to plant available P (Gahoonia & Nielsen, 1992).

The literature on soil P_o has been reviewed by Anderson (1975, 1980) and Dalal (1977), and labile P_o forms have been related to a number of soil forming factors by Harrison (1987). The complete P cycle was reviewed by Tate (1985),

Stevenson (1986), and Sanyal and De Datta (1992), and Stewart and Tiessen (1987) have presented a conceptual model of soil P_o turnover.

LONG-TERM RESEARCH SITES AND ORGANIC PHOSPORUS AVAILABILITY

Long-term budgets of soil organic matter for land under cultivation have repeatedly shown mineralization losses of C with concomitant release of significant amounts of N and P. In the Great Plains area, long-term research sites have been used to assess the contribution of soil organic matter (SOM) to N and P nutrition of crops (Campbell et al., 1990; Cole & Heil, 1981). Comparison of native and cultivated land at sites with known cultivation history, such as long-term plots in western Canada or the USDA sites in the Great Plains (Haas et al., 1961), show that cultivation and cropping without fertilizer inputs have seriously depleted not only soil organic matter, but also mineralizable N (Janzen, 1987a,b) and P_o in the soil (Tiessen et al., 1982; Tiessen & Stewart, 1983; McKenzie et al., 1992a).

The rate of organic matter mineralization during the first 40 to 60 yr of cultivation of Great Plains' soils liberated sufficient P for crop export to prevent P-limiting conditions. For instance, Tiessen et al. (1983) showed that the total P exported by crops from a prairie soil was less than that released from mineralizing SOM during 60 yr of cultivation. But at the later stages of the mineralization process, only more resistant SOM fractions remained and release rates were too slow to fulfill crop demands. During the period of mineralization, surplus P_i was precipitated as apatite-like compounds whose low solubility apparently does not satisfy crop uptake rates. As a consequence, fertilizer use in the area has greatly increased during the past 20 yr. Additional P from fertilization added at the Agriculture Canada Lethbridge site after 55 yr under the same cultivation practices has either caused a build up of residual P as P_i or, where both N and P have been added, a build up of P_o (McKenzie et al., 1992a,b). Phosphorus stores associated with high SOM levels have imparted resilience to an agricultural system that has traditionally mined soil nutrients with limited inputs. The importance of P_o mineralization in providing adequate P_i has been well established for temperate soils (Stewart & Sharpley, 1987; Stewart & Tiessen, 1987), and it appears to be even more important in the tropics (Tiessen, 1991).

Relatively rapid mineralization of organic matter from native vegetation or crops under tropical conditions can liberate sufficient P for plant growth to obscure the role of P_i extracted by P tests. Adepetu and Corey (1976) related plant uptake directly to the decrease in organic P during the cropping season following clearing of bush fallow. Agboola and Oko (1976) showed P_o mineralization to occur after land clearing at such a rate that available P was increased after the first crop despite crop uptake. The mineralization of P_o is particularly critical for plant supply in soils with high P sorption. On a Brazilian Oxisol, 6 yr of cultivation with minimal fertilization resulted in reductions of C, N, and P_o by 30% (Tiessen et al., 1992). A concomitant increase in P_i did not sustain supplies of labile P required for acceptable crop production levels because of high P sorption. Detailed P fractionation confirmed that the decline in P fertility was not due to net export of P in the crop, but due to the mineralization of P_o, and subsequent transformation of the surplus P_i to unavailable forms.

SHORT-TERM DYNAMICS

The documented long-term changes in soil P_o have usually been large enough to be visible in measurements of total P_o or a large portion of P_o. Short-term changes will be more important for the prediction of the contribution to soil fertility during a season. Studies of such changes have relied on the selective extraction of the most active P_o fractions, either from the bulk soil or the rhizosphere.

An indication of the short-term turnover of P_o is the variation in specific P_o extracts over seasons. Dormaar (1972) found a winter maximum of P_o in an Ap horizon of a dark brown Chernozem. Sharpley (1985) similarly reported that a net 20 to 74 kg P_o ha^{-1} were mineralized between the highest P_o level observed in the winter and the following summer low. On the other hand, summer maxima for P_o, coupled with minima in P_i fractions were observed on light textured arable, grassland and pasture sites by Magid and Nielson (1992), and a similar pattern has been observed earlier by Garbouchev (1966). Biomass P has also been shown to vary substantially with season. This variation shows little direct correlation with microbial C (Tate et al., 1991; Buchanan & King, 1992) and thus may reflect dynamics of the P cycle rather than of the biomass. This is in agreement with laboratory incubations performed by Chauhan et al. (1979), who showed no significant regression for estimates of chloroform-labile P against C.

Transformations of P in the plant–microbe–soil system were illustrated by McLaughlin et al. (1988a,b,c) in a double labeling experiment using ^{32}P fertilizer and ^{33}P alfalfa (*Medicago* sp.) residues. Wheat plants grown in the presence of both P sources acquired P from both (5.4% of ^{33}P and 11.6% of ^{32}P). The microbial biomass incorporated predominantly alfalfa residue-derived P (28% of ^{33}P, and <5% of ^{32}P) (McLaughlin et al., 1988b). Mediated by these microbial transformations, 40% of the alfalfa-derived P, although initially present in inorganic form in the residue, was incorporated into soil P_o in only 7 d (McLaughlin et al., 1988c). After 95 d, 85% of the ^{32}P recovered in the soil was found in inorganic forms, while more than 50% of the alfalfa-derived ^{33}P was in organic forms. Experiments of this kind are rare because of the short half lives of ^{32}P and ^{33}P.

The soil microbial biomass may be regarded as the driving force of the P cycle by which materials are taken up, converted into new products, and subsequently released (Stewart & Tiessen, 1987; Stewart & Sharpley, 1987; Van Veen et al., 1984). Accordingly, the chemical nature of soil P_o is largely determined by microbial products, and the dynamics of soil microbial biomass is a key factor in understanding the short-term dynamics of P_o in different ecosystems (Fig. 8–1).

In addition to mediating the turnover of P_o, microbes may constitute a significant reservoir of P (Brookes et al., 1982; Hedley & Stewart, 1982; McLaughlin & Alston, 1986). Brookes et al. (1984) estimated microbial P contents in six arable soils to range between 6 and 24 kg P ha^{-1}, on average amounting to 3% of soil P_o. Eight grassland soils held between 18 and 101 kg P ha^{-1}, an average of 14% of total soil P_o, in microbial biomass. Assuming a biomass

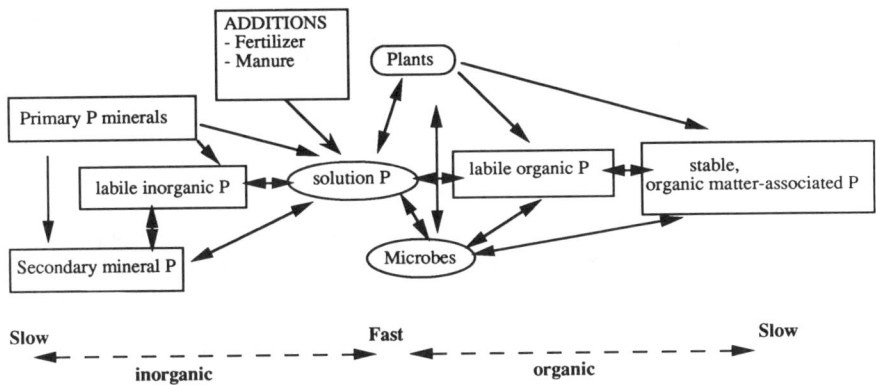

Fig. 8–1. A conceptual P cycle.

turnover rate of 0.4 per year, a maximal gross mineralization of 10 and 40 kg P ha^{-1} can be calculated for the arable and grassland soils, respectively (Brookes et al., 1984). This is equivalent to a substantial proportion of annual plant demands.

Most of the P associated with soil organisms is contained in bacteria and fungi. Higher levels of the trophic web, such as amoebae and nematodes contribute little to the P pool, but may play a key role in the net mineralization of P by accelerating the turnover of fungi and bacteria (Coleman et al., 1978; Elliott et al., 1980; Ryszkowski et al., 1989).

In biological or organic farming systems, the microbially mediated turnover of organic P may be more important than in conventional farming because manure is the only P material applied. Management of bio-dynamic and bio-organic farming systems aims towards a stimulation of biological soil activity (Besson & Niggli, 1991). Acid and alkaline phosphatase activity in the bulk soil of plots cultivated according to biological farming systems were higher than those seen in conventional fields (Oberson et al., 1993). The biological farming systems, therefore, had a higher potential to hydrolyze P_o. At the same time, the mineralization of organic C (Oberson, 1993) and the microbial biomass, determined as ATP (Oberson et al., 1993), were increased in the biologically cultivated plots. Such results indicate a higher turnover of organic substrates in the biological farming systems, in accordance with the greater importance of the external C input in form of manure into the biological systems. Available P determined by the ^{32}P isotopic exchange (Fardeau, 1993) and the comparison of P contents in the harvested plants (Oberson et al., 1993) showed that these biological–biochemical processes resulted in P availability with the organic system equal to that in exclusively minerally fertilized systems.

The mineralization of P_o may also be caused by free phosphatases derived from plants or microorganisms (McGill & Cole, 1981). Phosphatases exist as extracelluar or periplasmic hydrolases of soil microbes, and thus such abiotic mineralization of soil P_o may be closely tied to biological activity. Several studies show the importance of root phosphatase in the rhizosphere (Helal & Sauerbeck, 1984; Tarafdar & Jungk, 1987; Tarafdar & Claassen, 1988; Helal &

Dressler, 1989), which made soil P_o (Tarafdar & Jungk, 1987) and P_o added to soils (Helal & Dressler, 1989) available to plants. Phosphatase activity, however, did not determine the availability of P_o in the rhizosphere, because root phosphatase activity and P_o use were considerably reduced by a sufficient supply of P_i (enzymatic product inhibition). It was concluded that root phosphatase activity is a significant factor in nutrient uptake under limited P_i supply (Tarafdar & Claassen, 1988; Helal & Dressler, 1989).

The changes induced in soil adjacent to the growing plant root either through root exudation or root infection by mycorrhiza are an important aspect of the interrelationship between plants and microbial biomass. Many studies on the root rhizosphere (Hedley et al., 1982b; Kuchenbuch & Jungk, 1982; Farr et al., 1969) have documented that rhizosphere processes are important and that the pH of the rhizosphere may be changed. These studies illustrate that it is difficult to extract plant available P in a chemically well defined fraction, since fractions that are normally considered to be of low immediate plant availability (e.g., NaOH–P_o, nonextractable P) may contribute significantly to the nutrition of higher plants. Gahoonia and Nielsen (1992) found that 11 to 15% of the P depletion in the rhizosphere of rape (*Brassica napus L.*) seed was derived from P_o, and that >40% of the uptake was accounted for by fractions that are considered to be of low availability. An indirect role of P_o in the P turnover in the rhizosphere may be attributed to the competition of P_o for the same binding sites as P_i. Inositol hexaphosphate added to the soil as a soluble Na salt was sorbed rapidly with a concomitant increase of P_i concentration in the equilibrium solution (Dressler & Helal, 1992).

The mobilization of P by growing roots from fractions that are not readily accessible to isotopic exchange was also demonstrated using ^{32}P isotopic dilution (Helal & Dressler, 1989). A rapid and considerable redistribution of radioactivity among inorganic, organic, and microbial P in the rhizosphere due to microbial turnover processes was found. At the same time, the incorporation of P into the microorganisms from nonexchangeable fractions indicated the importance of microbial activity in the mobilization of P_i and P_o. Part of this effect may be due to the dissolution of normally insoluble P_i by microbes (Asea et al., 1988).

Mycorrhizae aid their host by exploring a soil volume outside the immediate rooting zone. Isotope labeling of soil indicates that the mycorrhizae deplete the same P sources as plants (Bolan, 1991; Morel & Plenchette, 1993). In addition, though, there may be effects on the biochemical availability of P_o in the rhizosphere since a 2- to 2.5-fold increase was found in the activity of acid phosphatase in the mycorrhizal rhizoplane of Norway spruce [*Picea abies* (L.) Karsten] in comparison to bulk soil (Häussling & Marschner, 1989). The use of organic P by the mycorrhizal roots was reflected by a decrease of water extractable P_o in the rhizoplane and also the rhizosphere soil. The rates of hydrolysis of organic P even exceeded the P-uptake rate by mycorrhizae and plant roots, resulting in an increase of P_i in the rhizosphere soil of spruce trees. The study of Häussling and Marschner (1989) is a rare example of the demonstration of the role of organic P and correspondingly of the phosphatase activity under field conditions.

In conclusion, both soil P_o and microbial P can be important P sources for plants. In addition, microorganisms may mobilize P_i of low solubility, which may then become plant available. The plant root also induces changes in the P turnover in the rhizosphere, for instance through the excretion of acid phosphatase and by pH changes. Seasonal changes in the biological activity of the soil and in the growth cycle of plants are therefore important components of the soil P cycle and will cause short-term variations in soils' available P status.

SPATIAL VARIABILITY

The importance of the rhizosphere and of microbial colonies in the turnover of soil P results in considerable spatial variability at a microscale. Accessibility of substrates and thus decomposition processes in terrestrial ecosystems vary between different pore sizes and across aggregates (Elliott et al., 1980; Van Veen & Kuikman, 1990). Only a fraction of the soil is biologically active, since <3% of the pore spaces are colonized by microorganisms, and many of these organisms are dormant. Soil extracts homogenize this important architecture, and yield averages that may not represent actual conditions.

The collection of soil water by expression or suction may allow discrimination between pore classes between macropores (>100 μm), mesopores (0.5–100 μm), or micropores (<0.5 μm) (Magid et al., 1992). Macropores are subject to leaching. Microbiological processes occur mainly in mesopores. Micropores are devoid of life, but stabilize biochemical products and interact with the biologically active soil through diffusion. Within larger pores access to microbes and predators affect P_o turnover (Cole et al., 1978), although such details are difficult to demonstrate in field soils.

Organic P concentrations in soil solution may be up to 20 times higher than those of P_i due to lower P_o adsorption to solid soil surfaces (Barel & Barsdate, 1978; Gersper et al., 1980), and it may be the main form in which P moves through soils (Hannapel et al., 1964; Campbell & Racz, 1975; Castro & Rolston, 1977; Hoffman & Rolston, 1980; Frossard et al., 1989). Schoenau and Bettany (1987) found that the proportion of P_o in the fulvic acid fraction increased with depth in soils developed under deciduous forest and native grassland, indicating a net downward movement of P_o. Magid and Nielsen (1992) showed distinct seasonal patterns of P_o movement at 90-cm depth, with winter minima during three consecutive years, indicating that effects from biological activity in the upper soil layers could be transmitted relatively rapidly to considerable depth.

At field scale, spatial variability in the distribution of P fractions and total P can be attributed to differences in parent materials, micro relief, soil moisture, clay content, plant uptake, and the redistribution of cations with which P interacts in profiles and landscapes (O'Halloran et al., 1986; Tiessen & Santos, 1989; Tiessen et al., 1991; Ball-Coelho et al., 1993). Coefficients of variation for labile P fractions may be in excess of 50% over only a few meters.

Cultivation, by its homogenizing action on the soil may reduce spatial variability from that found on native sites, while increased erosion rates may increase variations through selective removal of fine or light materials (Tiessen & Santos, 1989). Therefore, not only the presence of variability, but also its potential change must be taken into account when analyzing long-term temporal patterns in soil P distribution and cycling in the field.

METHODS FOR STUDYING ORGANIC PHOSPHORUS AND ITS TRANSFORMATIONS

Phosphate released into the soil solution rapidly interacts with the soils solid phase, minerals, and organic matter. Therefore, it is not possible to study P_o transformations and availability by incubation and subsequent leaching analogous to the mineralizable N method (Stanford & Smith, 1972). Chemical extractants have therefore been employed to characterize biochemical and biological stability, and the relationships between extractability and bioavailability have been inferred empirically.

Much effort has been expended on the chemical characterization soil P_o although the specific chemical form of P_o is rarely an indicator bio-availability. Such characterization requires the extraction of P_o, followed by often complex and tedious analyses for specific compounds such as nucleotides (Anderson, 1970), inositols (Cosgrove, 1980), lipids (Hance & Anderson, 1963), or others (Anderson, 1980; Halstead & Anderson, 1970). The interpretation of results was often limited by the uncertainty whether the extracted material was removed from the soil organic matter, microbes, or recent organic materials, and little further work has been done since the seventies.

Nuclear magnetic resonance (NMR) spectroscopy of ^{31}P can identify broad classes of P_o such as monoesters, diesters, or phosphonates (Wilson, 1987; Newman & Tate, 1980; Tate & Newman, 1982). Diesters are generally more labile in the soil environment and contents have been shown to decrease with long-term cultivation (Hawkes et al., 1984; Condron et al., 1990a). Monoesters make up most of the soil P_o and were shown to accumulate in pasture soils as a result of long-term P fertilizer additions (Hawkes et al., 1984; Condron et al., 1985). These results were all obtained on soil extracts, and the usefulness of liquid state ^{31}P NMR analysis of soil extracts for studying P_o dynamics, hinges on the skill with which suitable extracting agents are chosen to extract labile soil P_o (Adams & Byrne, 1989; Adams, 1990). Solid-state ^{31}P NMR must await further technical advances before it can be used to monitor quantitative changes in different forms of soil P_o.

An alternative approach to studying specific chemical forms of soil P_o has been the characterization of stability or bioavailability of P_o by extractions that discriminate different P_o fractions on the basis of the kind and strength of physicochemical interactions of P_o with other soil components (Harrison, 1982). Only a small portion of P_o participates in biological processes at any one time, and extractions have aimed at isolating bioavailable P. Two commonly employed alkaline extractants are 0.5 M sodium bicarbonate at pH 8.5 (analogous to the Olsen soil test), and varying strengths of sodium hydroxide, which is similar to the extractant for Fe- and Al-associated P_i (Chang & Jackson, 1957), or the commonly used extractants for soil organic matter. Bicarbonate-extractable P_o has been shown to be labile and available to plants and microbes in laboratory incubations (Bowman & Cole, 1978a) and comparative field studies (Tiessen et al., 1984). Hydroxide P_o is more stable and turns over more slowly in the field (Bowman & Cole, 1978a,b), although it can serve as a source for microbial P uptake during short-term incubations if P_i is limited (Chauhan et al., 1979).

Sample treatment can have critical effects on the results of such extractions. The amounts of P_o extracted with bicarbonate or hydroxide from soil kept as a laboratory standard by Magid and Nielsen (1992) showed coefficients of variation of >20%, and were very sensitive to temperature during the extraction.

Each of these extracts removes only one vaguely defined fraction of P_o, and the wish to describe the complexity of the total P_o cycle has led to the combination of several such extracts in sequential procedures. Kelley et al. (1983) found P_o in all fractions obtained with the sequential extraction of Chang and Jackson (1957). Labile fractions contained between 35 and 85% P_o. Hedley et al. (1982a) quantified chemically labile and more stable forms of P_o using a sequence of bicarbonate and hydroxide. The residue left after these alkaline extractions was analyzed in Ca-dominated soils by Tiessen et al. (1983) using a peroxide digest, and by Condron et al. (1990b), using a second NaOH treatment following a mild acid extraction. In more weathered soils, the residue was further extracted with hot acid (Condron et al., 1990b) similar to the procedure of Mehta et al. (1990).

The variation of P compositions of different soils of known history allowed an empirical interpretation of the roles of bicarbonate and hydroxide extractable P_o in P transformations associated with microbial P uptake (Hedley et al., 1982a), plant roots (Hedley et al., 1982b), cultivation (Tiessen et al., 1983), or soil development (Tiessen et al., 1984; Roberts et al., 1985). Using a path analysis on P fractions from 168 USDA benchmark soils from eight soil orders of the soil taxonomy system, Tiessen et al. (1984) showed that extractable P_o fractions accounted for a significant portion of the variability in plant available P in more weathered soils.

Despite these successes, the nature of the P_o extracts from the sequential fractionation is less well-defined than that of the inorganic extracts (Stewart & Tiessen, 1987). Much evidence from temperate soils suggests that bicarbonate P_o is the most labile fraction in the short-term. In long-term cultivation (60 yr), both bicarbonate P_o and hydroxide P_o were depleted equally (Tiessen et al., 1983). Thus relative short-term stability does not imply similar stability relationships in the long-term. Oberson et al. (1993) found no effect of biological and conventional farming systems on bicarbonate P_o after 14 yr of different cultivation. This result was interpreted as an indication of a rapid turnover of this labile fraction in the more biologically active systems where manure is applied. The bio-dynamic farming system, however, led to a change in the NaOH–P_o and even in residual, nonextractable P, which consisted mainly of P_o. Hence, organic fractions that are not easily soluble in chemical extractants, should be considered in transformations affecting P_o availability.

In a semiarid tropical Oxisol, ~6 yr of cultivation reduced hydroxide P_o by 30%, while bicarbonate P_o levels remained constantly low (Tiessen et al., 1992). Similarly, the build-up of P_o as a result of fertilizer additions was only reflected in the hydroxide P_o fraction of this tropical soil. Thus, the bicarbonate P_o pool appears to be at a constant, minimal level regardless of the cropping history in some tropical soils. In contrast, the hydroxide P_o pool reflects the overall changes in soil organic matter and P_o levels when the soil is stressed by cultivation and net P export. This pool may therefore represent a relatively active reservoir (source or sink) of P under tropical conditions (Tiessen et al., 1992).

In several cases, the residual P of the original Hedley et al. (1982a) fractionation also showed relatively short-term transformations despite its recalcitrance with respect to chemical extractants. Part of this residual P can be identified as P_o upon hot acid extraction (Tiessen & Moir, 1983). A possible explanation for the bioavailability is that it is P_o held in organic debris protected by (hot HCl hydrolyzable) cellulose, and easily liberated during C turnover.

Soil solution composition, and exchangeable Ca can affect the distribution of soil P_i and P_o over different fractions (Perrot, 1992). Liming of a soil can thus have pronounced effects on the apparent P_o composition, even in short-term treatments, of insufficient time to effect true P_o transformations (Condron et al., 1993).

In summary, fractions that are labile in the long-term may not be labile in the short-term and what is liable in the tropics may be stable in temperate soils. The terms labile, stable and so on are associated with specific extracts and must, therefore, be qualified in specific situations and cannot be universally applied.

Some P_o fractions have been found to change predictably under manipulations such as incubations, fertilizations, manuring, or exhaustive cropping. To fully understand the P_o cycle, though organic matter transformations must be considered (Tiessen, 1991). Many active biological structures are resistant to extraction (Stevenson & Elliot, 1989), and therefore some nonextractable P_o is biologically active as a result of organic matter turnover. Separation of recent litter, floatable organic matter, clay associated or mineral stabilized organic matter (Feller et al., 1983), and analysis of associated P (Tiessen et al., 1983) are methods that have shown some success, since they separate organic matter fractions of different stability. The determination of the C/N/P ratio in different extracts might add some information about the nature and the availability of the organic P extracted. The interpretation of these results, however, demands a combination with other chemical soil characteristics, and the extensive analytical work required is often prohibitive.

Microbial P may be a source of available P, and its fluctuations may be interpretable in terms of the soil P cycle. Chloroform fumigation of soil, based on the method for biomass determination by Jenkinson and Powlson (1976), has been widely accepted as a means for rendering microbially held P extractable with mild reagents. The recovery of microbial P by fumigation is always low, since cell lysis is known to be incomplete, and upon release, P reacts with other soil components. Correction factors are routinely used to arrive at microbial P measures (Brookes et al., 1982; Hedley & Stewart, 1982; McLaughlin & Alston, 1986). Fresh organic materials or roots may increase apparent microbial P values (McLaughlin & Alston, 1986; Sampaio et al., 1986), although Badalucco et al. (1992) reported, based on biochemical characterization of soil organic compounds extracted with K_2SO_4 (0.5 M) before and after fumigation, that fumigation did not increase the release of nonbiomass C and N compounds.

Increases in $NaHCO_3$ extractable P upon drying of soils have been attributed to the death of microorganisms during the drying process (Brookes et al., 1982; Sparling et al., 1985). Sparling et al. (1987) found an effect of predominant soil moisture regime on the microbial contribution to $NaHCO_3$ extractable P upon drying, presumably due to microbial adaptation to soil moisture conditions.

Cowling et al. (1987) noted that aging of the $NaHCO_3$ extract used in the extraction of microbial P may cause some variability in extraction results. Field history and subsequent sample treatment can thus affect the results of microbial P determinations.

Only the quantity of microbial biomass P can be estimated by fumigation, and fluxes and activities of specific groups of organisms cannot be shown. The P content in different fungi and bacteria varied from 0.4 to 2.7% P (oven-dry basis; Brookes et al., 1982) with higher contents in bacteria than in fungi. Specific soil invertebrates have been isolated using flotation and centrifugation, and their P contents determined (McKercher et al., 1979). Although the proportion of total P_o in these organisms is small, they play an important role in P fluxes. The turnover rate of the biomass is affected by cultivation practices such as the application of farmyard manure (Goyal et al., 1993) and crop rotation (McGill et al., 1986), which affects the annual P flux through the microbial biomass as calculated by Brookes et al. (1984). It may, therefore, be useful to employ a measure of microbial P that also gives an indication of activity. Oberson (1993) calculated microbial P based on adenosine triphosphate (ATP) determinations (Maire, 1984). ATP is a good indicator of the size of the microbial biomass (Witter et al., 1993), and also a critical metabolite and labile organic P compound. As the ratio (ATP content of the soil–biomass C content of the soil) is relatively constant (Jenkinson et al., 1979), temporal fluctuations of the ATP content may give information about P flux through the biomass.

Attempts to use ^{32}P labeling for the study of specific P_o transformations have generally not been successful because of the multiple reactions that added ^{32}P undergoes. It has been argued that mineralization of P_o will result in a dilution of added $^{32}PO_4$, and therefore, it has been suggested that differences in dilution rates may have the potential of yielding information on P_o mineralization rates, but several concurrent processes are involved in isotope dilution. It is well documented that the reaction of P_i with the mineral phase is a continuous process that will only slowly reach equilibrium in soil (Fardeau & Jappé, 1976). The microbial biomass may also take up the $^{32}PO_4$ as a result of the addition of organic matter, or due to changes in physiological states caused by moisture fluctuations or other disturbances. Thus, the use of isotope dilution as a measure of mineralization of P_o requires a control of both the ^{32}P content in soil microbial biomass as well as the reactions of $^{32}PO_4$ with the soil. The size of the exchangeable P_i pool is such, that relatively small isotopic dilution effects from the mineralization of P_o are likely to be lost in the margins of error.

An alternative approach was taken by Walbridge and Vitousek (1987) who attempted to label the soil homogeneously with ^{32}P in the hope that soil P_o would be the only fraction that remained unlabeled, because of its slow turnover. Mineralization of soil P_o would then cause dilution of ^{32}P. Release of ^{32}P from microbes and microbially mediated labeling of P_o has limited the success of this approach. Labeled materials have been successfully added to soil in order to follow decomposition of plant material (McLaughlin et al., 1988a) or to follow the turnover of labile substrates in soils (Harrison, 1982), but the short half lives of both P isotopes limit the useful period of observation to less than a season. The high initial activities needed normally prevent any field applications.

It is interesting to compare the progress made in integrating organic matter and soil fertility management concepts along with other production factors into production agriculture. At a symposium sponsored by the Soil Science Society of America in December, 1985, the statement was made that computer simulation is a significant innovation and is expected to have a major impact on farming practices. The development of our knowledge of processes governing plant growth, soil organic matter formation, mineralization, and other production factors has progressed to the extent that they can be described mathematically in simulation models. These models are interactive, and a couple are controlled by environmental factors, such as the availability of soil water, photosynthetic radiation, and climatic variables. Simulation of the processes involved can be linked to existing data bases to provide a powerful predictive capability to aid farm operators in management and decision making (Stewart et al., 1987).

This statement applies, as well, to our ability to predict available P in soil. It is possible to model some of the effects of crop management and rotation systems including cultivation practices in determining the speed at which phosphorus will cycle within the soil system (Parton et al., 1987). For instance the CENTURY model, which was just described in 1983, has now reached CENTURY Agroecosystem Version 4.0 and this version has been thoroughly tested and adapted to predict changes in management and climate on ecosystems (Metherell et al., 1993). Two figures are reproduced from this model; the first (Fig. 8–2) shows the data that goes into the File Manager part of the model, and the second (Fig. 8–3) shows the type of detail in the P sub model. This model has the capability of predicting available (active) P_o under most North American climate and management systems.

Soil testing laboratories have developed some sophistication in accessing soil data bases, climatic conditions, and predicting fertilizer response with a fair degree of precision, especially under practices such as the wheat fallow system in Great Plains soils. We also have the capability of understanding how rotation systems, different methods of cultivation and trash management will affect the levels of organic C, mineralizable N, and P in soils. These models have been developed using some of the long-term sites and have been tested at other sites across the Great Plains area. At the present time, such simulation models could be linked to soil test models to provide a more accurate assessment of soil fertility. This has been attempted with a degree of success in experimental conditions, but a step of faith is required at the present time for these to be incorporated into commercial systems. Application of models is not yet possible across all soil types. The factors that stabilize organic matter in soils are altered by the degree of leaching and weathering and other pedogenic processes, and an improved understanding of organic matter and nutrient cycling in tropical soils is required before models can be applied to the tropics.

INNOVATIVE AVAILABILITY INDICES ASSESSING ORGANIC PHOSPHORUS

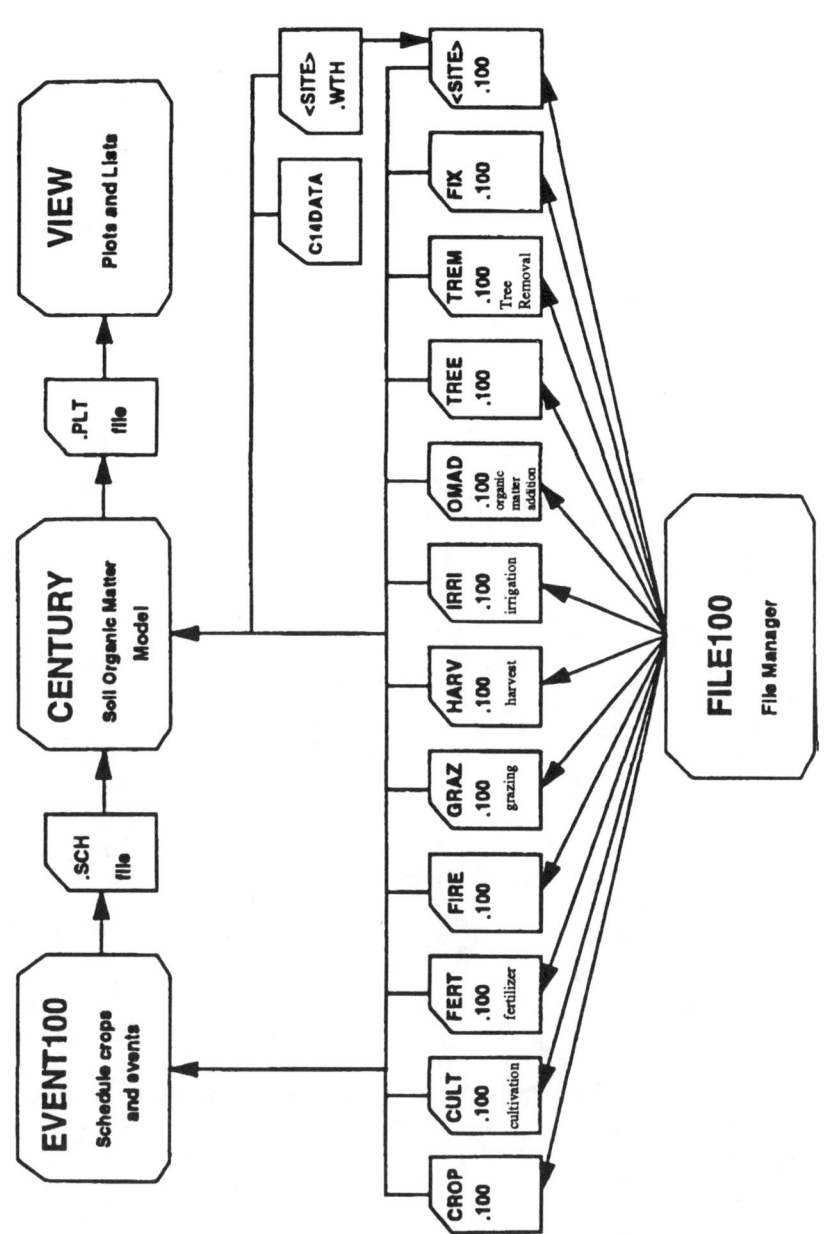

Fig. 8–2. Details of inputs into CENTURY—File Manager.

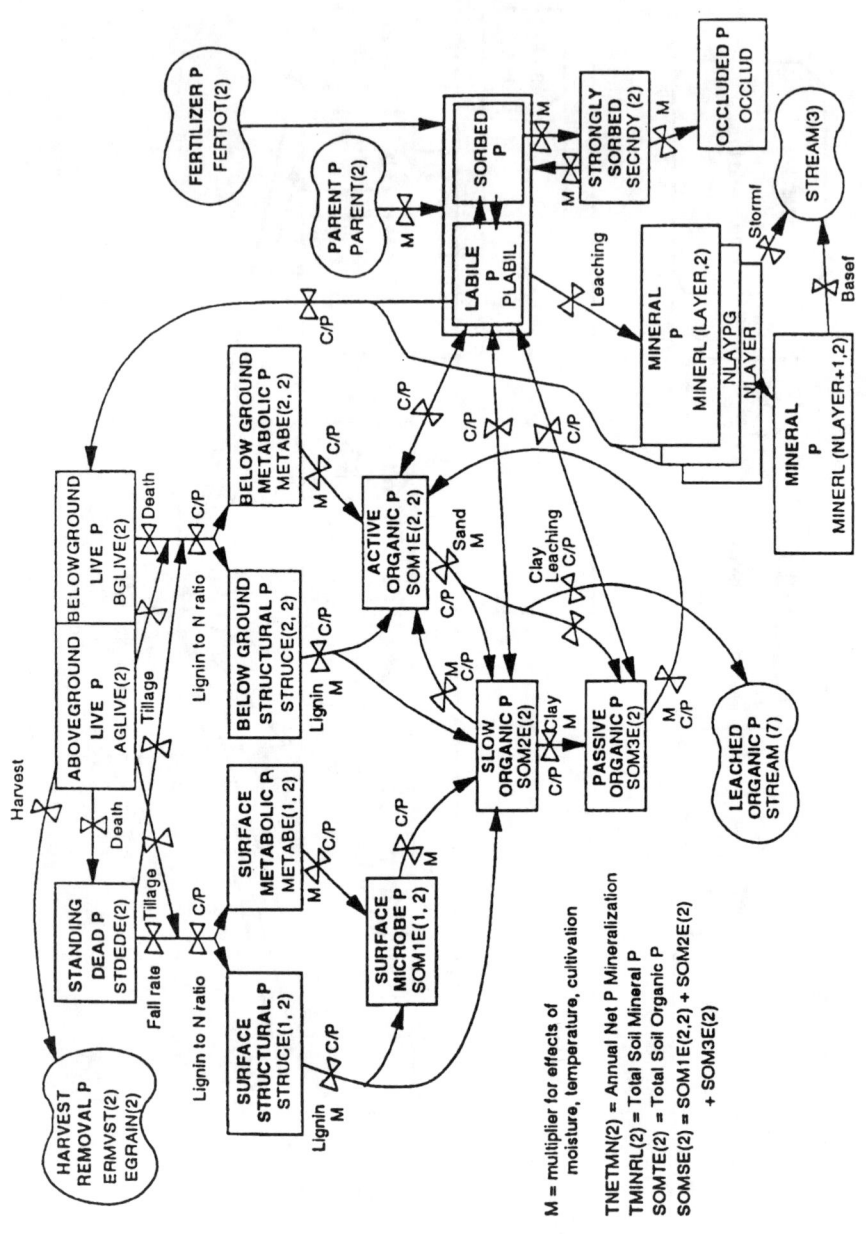

Fig. 8–3. Details of the P submodel in CENTURY.

CONCLUSIONS

Studies of the interactions of C, N, P, and S in soil organic matter and their contribution to plant nutrition and crop production in several natural and managed systems have led to the development of concepts of organic matter and P_o turnover. An important tool for testing ideas and concepts on P_o transformations is the modeling of organic matter components (Parton et al., 1987) and associated P (Woomer, 1992), which may provide a tool for comparing extractable fractions with (hypothetical) functional pools. These studies have allowed us to develop concepts and procedures in quantitative pedology and to test these concepts on soils selected along climo-, topo-, and cultivation chronosequences (Stewart & Cole, 1989). Initially, we concentrated on soils developed on the former grasslands and bordering forest soils of the Great Plains of North America, but more recently these studies have been expanded to include more weathered soils that are found in temperate and tropical conditions (Tiessen et al., 1984, 1992).

Despite recent advances, a great deal remains to be understood about the nature and dynamics of soil P_o. One of the critical aims is the identification of biologically meaningful P_o pools in the soil. Their identification would provide tools for the study of the fate of P_o in soil and the influence of various anthropogenic and environmental factors. This will require continued development of methodology designed to identify and characterize spatially and physically distinct forms of P_o in soil, which can be used to examine short- and long-term transformations of P_o. There is also a need to continue investigations into P_o dynamics associated with the soil microbial biomass, mechanisms of P_o mineralization in soil (and the role of phosphatase enzymes in particular), and relationships between the cycling of P_o and other soil organic constituents and nutrients.

The complex, interactive nature of the various factors and processes which influence P_o cycling in soil requires the development and refinement of appropriate simulation models, especially over the short-term. Further development of simulation models would also be valuable in directing future research. Thus, it is not possible to be optimistic that a rapid soil test method is forthcoming in the near future that provides a cheap, rapid test for available P. However, it is becoming very important that soil tests that adequately determine available PO_4 in soils be supplemented with a simulation of P_o turnover under specific management practices. This will allow reliable estimates of inputs (kg P ha^{-1}) of P_o to the labile P soil pool. Organic matter models have reached a degree of sophistication that permits adequate prediction in many soil types (Parton et al., 1987). In the meantime, secondary indicators of biological activity, organic matter fractions, litter return, and others may provide evidence that could be used to predict P_o availability in the short-term.

REFERENCES

Adams, M.R. 1990. ^{31}P NMR identification of phosphorus compounds in neutral extracts of mountain ash (*Eucalyptus regnans* F. Meull) soils. Soil Biol. Biochem. 22:419–421.

Adams, M.A., and L.T Byrne. 1989. ^{31}P NMR analysis of phosphorus compounds in neutral extracts of surface soils from selected Karri (*Eucalyptus diversicolor* F. Muell) forests. Soil Biol. Biochem. 21:525–528.

Adepetu, J.A., and R.B. Corey. 1976. Organic phosphorous as a predictor of plant-available phosphorus in soils of southern Nigeria. Soil Sci. 122:159–164.

Agboola, A.A., and B. Oko. 1976. An attempt to evaluate plant available P in western Nigerian soils under shifting cultivation. Agron. J. 68:798–801.

Anderson, G. 1970. The isolation of nucleoside diphosphates from alkaline extracts of soil. J. Soil Sci. 21:96–104.

Anderson, G. 1975. Other organic phosphorus compounds. p. 295–331. In J.E. Geiseking (ed.) Soil components: Organic components. Vol. 1. Springer, Berlin.

Anderson, G. 1980. Assessing organic phosphorus in soils. p. 411–428. In F.E. Khasawneh (ed.) The role of phosphorus in agriculture. ASA, CSSA, and SSSA, Madison, WI.

Anderson, G., E.G Williams, and J.O. Moir. 1974. A comparison of the sorption of inorganic orthophosphate and inositol hexaphosphate by six acid soils. J. Soil Sci. 25:51–62.

Asea, P.E.A., R.M.N. Kucey, and J.W.B. Stewart. 1988. Inorganic phosphate solubilization by two Penicillium species in solution culture and soil. Soil Biol. Biochem. 20:459–464.

Badalucco, L., A. Gelsomino, S. Dell'Orco, S. Grego, and P. Nannipieri. 1992. Biochemical characterization of soil organic compounds extracted by 0.5 M K_2SO_4 before and after chloroform fumigation. Soil Bio. Biochem. 24:569–578.

Ball-Coelho, B., I.H. Salcedo, H. Tiessen, and J.W.B. Stewart. 1993. Short- and long-term dynamics in a fertilized Ultisol cultivated with sugar cane. Soil Sci. Soc. Am. J. 57:1027–1034.

Barel, D., and R.J. Barsdate. 1978. Phosphorus dynamics of wet coastal Tundra soils near Barrow, Alaska. In Environmental chemistry and cycling processes. Proc. Symp. Augusta, GA. 28 Apr.–1 May 1976. Technical Information Center. U.S. Dep. of Energy, Washington, DC.

Batsula, A.A., and G.M. Krivonosova. 1973. Phosphorus in the humic and fulvic acids of some Ukrainian soils. Sov. Soil Sci. 5:347–350.

Beever, R.E., and D.J.W. Burns. 1980. Phosphorus uptake, storage and utilization by fungi. Adv. Bot. Res. 8:128–219.

Besson, J.M., and U. Niggli. 1991. DOK-Versuch: vergleichende LangzeitUntersuchungen in den drei Anbausystemen biologisch-dynamisch, organischbiologisch und konventionell: 1. Konzeption des DOK-Versuches: 1. und 2. Fruchtfolgeperiode. Schweiz. Landwirtsch. Forsch. 31:79–109.

Bieleski, R.L. 1973. Phosphate pools, phosphate transport and phosphate availability. Annu. Rev. Plant Physiol. 24:225–252.

Bolan, N.S. 1991. A critical review on the role of mycorrhizal fungi in the uptake of phosphorus by plants. Plant Soil. 134:189–207.

Bowman R.A., and C.V. Cole. 1978a. An exploratory method for fractionation of organic phosphorus from grassland soils. Soil Sci. 25:95–101.

Bowman, R.A., and C.V. Cole. 1978b. Transformation of organic phosphorus substrates in soils as evaluated by $NAHCO_3$-extraction. Soil Sci. 125:49–54.

Brookes, P.C., D.S. Powlson, and D.S. Jenkinson. 1982. Measurement of microbial biomass phosphorus in soil. Soil Biol. Biochem. 14:319–329.

Brookes, P.C., D.S. Powlson, and D.S. Jenkinson. 1984. Phosphorus in the soil microbial biomass. Soil Biol. Biochem. 16:169–175.

Buchanan, M., and L.D. King. 1992. Seasonal fluctuations in soil microbial biomass carbon, phosphorus and activity in no-till and reduced-chemical-input maize agroecosystems. Bio. Fertil. Soils. 13:211–217.

Campbell, C.A., R.P. Zentner, H.H. Janzen, and K.C. Bowren, 1990. Crop rotation studies on the Canadian prairies. Publ. 1841/E. Agriculture Canada Research Branch, Ottawa, Ontario.

Campbell, L.B., and G.J. Racz. 1975. Organic and P content, movement and mineralization of P in soil beneath a feedlot. Can. J. Soil Sci. 55:457–466.

Castro, C.L., and D.E. Rolston. 1977. Organic phosphate transport and hydrolysis in soil: Theoretical and experimental evaluation. Soil Sci. Soc. Am. J. 41:1085–1092.

Chang, S.C., and M.L. Jackson. 1957. Fractionation of soil phosphorus. Soil Sci. 84:133–144.

Chauhan, B.S., J.W.B. Stewart, and E.A. Paul. 1979. Effect of carbon additions on soil labile inorganic, organic and microbially held phosphate. Can. J. Soil Sci. 59:387–396.

Cole, C.V., E.T. Elliot, H.W. Hunt, and D.C. Coleman. 1978. Trophic interactions in soils as they affect energy and nutrient dynamics: V. Phosphorus transformations. Microb. Ecol. 4:381–387.

Cole, C.V., and R.D. Heil. 1981. Phosphorus effects on terrestrial nitrogen cycling. Ecol Bull. 33:363–374.

Coleman, D.C., R.V. Anderson, C.V. Cole, E.T. Elliot, L. Woods, and M.K. Campion. 1978. Trophic interactions in soils as they affect energy and nutrient dynamics: IV. Flows of metabolic and biomass carbon. Microb. Ecol. 4:373–380.

Condron, L.M, E. Frossard, H. Tiessen, R.H. Newman, and J.W.B. Stewart. 1990a. Chemical nature of organic phosphorus in cultivated and uncultivated soils under different environmental conditions. J. Soil Sci. 41:41–50.

Condron, L.M., K.M. Goh, and R.H. Newman. 1985. Nature and distribution of soil phosphorus as revealed by a sequential extraction method followed by ^{31}-P nuclear magnetic resonance analysis. J. Soil Sci. 36:199–207.

Condron, L.M., J.O. Moir, H. Tiessen, and J.W.B. Stewart. 1990b. Critical evaluation of methods for determining total organic phosphorus in tropical soils. Soil Sci. Soc. Am. J. 54:1261–1266.

Condron, L.M., C. Trasar-Cepeda, H. Tiessen, J.O. Moir, and J.W.B. Stewart. 1993. Effects of liming on organic matter decomposition and phosphorus extractability in an acid humic Ranker soil from northwestern Spain. Biol. Fertil. Soils. 15:279–284.

Cosgrove, D.J. 1980. Inositol phosphates. Studies in organic chemistry. Vol. 4. Elsevier Scientific Publ., Amsterdam.

Cowling, J.C., T.W. Speir, and H.J. Percival. 1987. Potential problems with the determination of Olsen and microbial P of soils due to the instability of 0.5 M sodium bicarbonate. Commun. Soil Sci. Plant Anal. 18:637–652.

Dalal, R.C. 1977. Soil organic phosphorus. Adv. Agron. 29:3–17.

Dormaar, J.F. 1972. Seasonal pattern of soil organic phosphorus. Can. J. Soil Sci. 52:107–112.

Dressler, A., and H.M. Helal. 1992. Interactions and mobility of phytate phosphorus in soil p. 602–603. *In* Proc. of the 4th Int. Conference, Gent. 8–11 Sept. 1992. Institut Mondial du Phosphate Casablanca, Morocco.

Elliot, E.T., R.V. Anderson, D.C. Coleman, and C.V. Cole. 1980. Habitable pore space and microbial trophic interactions. Oikos. 35:327–355.

Fardeau, J.C. 1993. Le phosphore assimilable du sol: Sa représentation par un modèle foncionnel a plusieurs compariments. Agronomie 13:317–331.

Fardeau, J.C., and J. Jappé. 1976. Nouvelle méthode de détermination du phosphore assimilable du sol par les plantes: Extrapolation des cinétiques de dilution isotopique. C.R. Acad. Sci. Ser. D. 282:1137–1140.

Farr, E., L.V. Vaidyanathan, and P.H. Nye. 1969. Measurement of ionic concentration gradients in soil near roots. Soil Sci. 102:385–391.

Feller, C., F. Bernhardt-Reversat, J.L. Garcia, J.J. Pantier, and S. Roussos. 1983. Études de la matière organique de différentes fractions granulométriques d'un sol sableux tropical. Effet d'un amendment organique (compost). Cah. ORSTOM Ser. Pedol. 20:223–238.

Frossard, E., J.W.B. Stewart, and R.J. St. Arnaud. 1989. Phosphorus distribution as related to its form and mobility in grassland and forest soils in Saskatchewan. Can. J. Soil Sci. 69:401–416.

Gahoonia, T.S., and N.E. Nielsen. 1992. The effects of root-induced pH changes on the depletion of inorganic and organic phosphorus in the rhizosphere. Plant Soil. 143:185–191.

Garbouchev, I.P. 1966. Changes occurring during a year in the soluble phosphorus and potassium in soil under crops in rotation experiments at Rothamsted, Woburn and Saxmundham. J. Agric. Sci. (Cambridge) 66:399–412.

Gersper, P.L., V. Alexander, S.A. Barkley, R.J. Barsdate, and P.S. Flint. 1980. The soils and their nutrients. p. 219–254. *In* An Arctic Ecosystem. The coastal tundra at Barrow, Alaska. Dowden, Hutchison & Ross, Stroudsburg, PA.

Goyal, S., M.M. Mishra, S.S. Dhankar, K.K. Kapor, and R. Batra. 1993. Microbial biomass turnover and enzyme activities following the application of farmyard manure to field soils with and without pervious long-term applications. Biol. Fertil. Soils. 15:60–64.

Haas, H.J., H.J. Grunes, and G.A. Reichmann. 1961. P changes in Great Plains soils as influenced by cropping and manure application. Soil Sci. Soc. Am. Proc. 25:214–218.

Halstead, R.L., and G. Anderson. 1970. Chromatographic fractionation of organic phosphates from alkali, acid and aqueous acetylacetone extracts of soils. Can. J. Soil Sci. 50:111–119.

Hance, R.J., and G. Anderson. 1963. Extraction and estimation of soil phospholipids. Soil Sci. 96:157–161.

Hannapel, R.J., W.H. Fuller, S. Bosma, and J.S. Bullock. 1964. Phosphorus movement in a calcareous soil: I. Predominanace of organic forms of phosphorus in phosphorus movement. Soil Sci. 97:350–357.

Harrison, A.F. 1982. Labile organic phosphorus mineralization in relationship to soil properties. Soil Biol. Biochem. 14:43–51.

Harrison, A.F. 1987. Soil organic phosphorus–a review of world literature. C.A.B. Int., Wallingsford, England.

Häussling, M., and H. Marschner. 1989. Organic and inorganic soil phosphates and acid phosphatase activity in the rhizosphere of 80-year-old Norway spruce [*Picea abies* (L.) Karst.] trees. Biol. Fertil. Soils 8:128–133.

Hawkes, G.E., D.S. Powlson, E.W. Randall, and K.R. Tate. 1984. A ^{31}-P nuclear magnetic resonance study of the phosphorus species in alkali extracts of soils from long-term field experiments.

Hedley, M.J., and J.W.B. Stewart. 1982. Method to measure microbial phosphate in soils. Soil Biol. Biochem. 14:377–385.

Hedley, M.J., J.W.B. Stewart, and B.S. Chauchan. 1982a. Changes in inorganic and organic soil phosphorus fractions induced by cultivation practices and laboratory incubations. Soil Sci. Soc. Am. J. 46:970–976.

Hedley, M.J., R.E. White, and P.H. Nye. 1982b. Plant induced changes in the rhizosphere of rape (*Brassica napus* var. *Emerald*) seedlings: III. Changes in L value, soil phosphate fractions and phosphatase activity. New Phytol. 91:5–6.

Helal, H.M., and A. Dressler. 1989. Mobilization and turnover of soil P in the rhizosphere. Z. Pflanzenernaehr. Bodenkd. 152:175–180.

Helal, H.M., and D. Sauerbeck. 1984. Influence of plant roots on C and P metabolism in soil. Plant Soil. 76:175–182.

Hoffman, D.L., and D.E. Rolston. 1980. Transport of organic phosphate in soil as affected by soil type. Soil Sci. Soc. Am. J. 44:46–52.

Janzen, H.H. 1987a. Effect of fertilizer on soil productivity in long-term wheat rotations. Can. J. Soil Sci. 67:165–174.

Janzen, H.H. 1987b. Soil organic matter characteristics after long-term cropping to various spring wheat rotations. Can. J. Soil Sci. 67:845–856.

Jenkinson, D.S., S.A. Davidson, and D.S. Powlson. 1979. Adenosine triphosphate and microbial biomass in soil. Soil Biol. Biochem. 11:521–527.

Jenkinson, D.S., and D.S. Powlson. 1976. The effects of biocidal treatments on metabolism in soil. V. A method of measuring soil biomass. Soil Biol. Biochem. 8:209–213.

Kelley, J., M.J. Lambert, and J. Turner. 1983. Available phosphorus forms in forest soils and their possible ecological significance. Commun. Soil. Sci. Plant Anal. 14:1217–1234.

Kowalenko, C.G. 1978. Organic nitrogen, phosphorus and sulfur in soils. p. 95–136. *In* Soil organic matter. Elsevier, New York.

Kuchenbuch, R., and A. Jungk. 1982. A method for determining concentration profiles at the soil root-interface by thin slicing rhizospheric soil. Plant Soil. 68:91–94.

Magid, J., N. Christensen, and H. Nielsen. 1992. Measuring phosphorus fluxes through the root zone of a layered sandy soil: Comparisons between lysimeter and suction cell solution. J. Soil Sci. 43:739–747.

Magid, J., and N.E. Nielsen. 1992. Seasonal variation in organic and inorganic phosphorus fractions of temperate climate sandy soil. Plant Soil. 144:155–165.

Maire, N. 1984. Extraction de l'adenosine dans les sols: Une nouvelle méthode de call des partes e ATP. Soil Biol. Biochem. 16:361–366.

McGill, W.B., K.R. Cannon, J.A. Robertson, F.D. Cook. 1986. Dynamics of soil microbial biomass and water-soluble organic C in Breton L after 50 years of cropping to two rotations. Can. J. Soil Sci. 66:1–19.

McGill, W.B., and C.V. Cole. 1981. Comparative aspects of cycling of organic C, N, S, and P through soil organic matter. Geoderma. 26:267–286.

McKenzie, R.H., J.W.B. Stewart, J.F. Dormaar, and G.B. Schaalje. 1992a. Long-term crop rotation and fertilizer effects on phosphorus transformations: I. In a Chernozemic soil. Can. J. Soil Sci. 72:569–579.

McKenzie, R.H., J.W.B. Stewart, J.F. Dormaar, and G.B. Schaalje. 1992b. Long-term crop rotation and fertilizer effects on phosphorus transformations: II. In a Luvisolic soil. Can. J. Soil Sci. 72:581–589.

McKercher, R.B, T.S. Tollefson, and J.R. Willard. 1979. Biomass and phosphorus contents of some soil invertebrates. Soil Biol. Biochem. 11:387–391.

McLaughlin, M.J., and A.M. Alston. 1986. Measurement of phosphorus in the soil microbial biomass: A modified procedure for field soils. Soil Bio. Biochem. 18:437–443.

McLaughlin, M.J., A.M. Alston, and J.K. Martin. 1988a. Phosphorus cycling in wheat-pasture rotations: I. The source of phosphorus taken up by wheat. Aust. J. Soil Res. 26:323–331.

McLaughlin, M.J., A.M. Alston, and J.K. Martin. 1988b. Phosphorus cycling in wheat-pasture rotations: II. The role of the microbial biomass in phosphorus cycling. Aust. J. Soil Res. 26:333–342.

McLaughlin, M.J., A.M. Alston, and J.K. Martin. 1988c. Phosphorus cycling in wheat-pasture rotations: III. Organic phosphorus turnover and phosphorus cycling. Aust. J. Soil Res. 26:343–353.

Mehta, N.C., J.O. Legg, C.A.I. Goring, and C.A. Black. 1954. Determination of organic phosphorus in soils: I. Extraction method. Soil Sci. Soc. Am. Proc. 18:443–449.

Metherell, A.K., L.A. Harding, C.V. Cole, and W.J. Parton. 1993. CENTURY Soil Organic Model Environment Technical Documentation Agroecosystem Version 4.0. Great Plains System Research Unit Tech. Rep. 4.0. USDA-ARS, Fort Collins, CO.

Morel, C., and C. Plenchette. 1993. Is the isotopically exchangable phosphorus in loamy soil the plant available phosphorus? Plant Soil 158:287–297.

Newman, R.H., and K.R. Tate. 1980. Soil characterized by ^{31}-P nuclear magnetic resonance. Commun. Soil Sci. Plant Anal. 11:835–842.

Oberson, A. 1993. Phosphordynamik in biologisch und knoventionell bewirtschafteten Böden. Ph.D. diss. Swiss Federal Insitute of Technology, Zürich (Diss. Abstr. Nr. 10119).

Oberson, A., J.C. Fardeau, J.M. Besson, and H. Sticher. 1993. Soil phosphorus dynamics in cropping systems managed according to conventional and biological agricultural methods. Biol. Fertil. Soils 16:111–117.

O'Halloran, I.P., R.G. Kachanoski, and J.W.B. Stewart. 1986. Influence of the spatial distribution of sand content on sampling patterns. Can. J. Soil Sci. 66:641–652.

Parton, W.J., D.S. Schimel, C.V. Cole, and D.S. Ojima. 1987. Analysis of factors controlling soil organic matter levels in Great Plains grasslands. Soil Sci. Soc. Am. J. 51:173–179.

Perrot, K.W. 1992. Effects of exchangeable calcium on fractionation of inorganic and organic soil phosphorus. Commun. Soil Sci. Plant Anal. 23:827–840.

Roberts, T.L., J.W.B. Stewart, and J.R. Bettany. 1985. The influence of topography on the distribution of organic and inorganic soil phosphorus across a narrow environmental gradient. Can. J. Soil Sci. 65:651–665.

Ryszkowski, L., J. Karg, B. Szpakowska, and I. Zyczynska-Baloniak. 1989. Distribution of phosphorus in meadow and cultivated field ecosystems. p. 178–192. *In* Phosphorus cycles in terrestrial and aquatic ecosystems. H. Tiessen (ed.) Saskatchewan Inst. of Pedology, Saskatoon.

Sampaio, E.V.S.B., I.H. Salcedo, and L.C. Maia. 1986. Limitaçoes no cálculo da biomassa microbiana determinada pelo método da fumigaçao em solos com adiçao recente de substrato organico (14C). Rev. Bras. Cienc. Solo 10:31–35.

Sanyal, S.K., and De Datta. 1992. Chemistry of phosphorus transformations in soil. Adv. Soil Sci. 16:1–120.

Schoenau, J.J., and J.R. Bettany. 1987. Organic matter leaching as a component of carbon, nitrogen, phosphorus and sulfur cycling in a forest, grassland and gleyed soil. Soil Sci. Soc. Am. J. 51:646–651.

Shang, C., P.M. Huang, and J.W.B. Stewart. 1990. Kinetics of adsorption or organic and inorganic phosphates by short-range ordered aluminum precipitate. Can. J. Soil Sci. 70:461–470.

Shang, C., P.M. Huang, and J.W.B. Stewart. 1992. pH effect on kinetics of adsorption of organic and inorganic phosphates by short-range ordered aluminum and iron precipitates. Geoderma 53:1–14.

Sharpley, A.N. 1985. Phosphorus cycling in unfertilized and fertilized agricultural soils. Soil Sci. Soc. Am. J. 49:905–911.

Sparling, G.P., J.D.G. Milne, and K.W. Vincent. 1987. Effect of soil moisture regime on the microbial contribution to Olsen P values. N.Z. J. Agric. Res. 30:79–84.

Sparling, G.P., K.N. Whale, and A.J. Ramsay. 1985. Quantifying the contribution from the soil microbial biomass to the extractable P levels of fresh and air-dried soils. Aust. J. Soil Res. 23:613–621.

Stanford, G., and S.J. Smith. 1972. Nitrogen mineralization potentials of soils. Soil Sci. Soc. Am. Proc. 36:465–472.

Stevenson, F.J. 1986. The phosphorus cycle. p. 231–284. *In* Cycles of soil carbon, nitrogen, phosphorus, sulfur and micronutrients. John Wiley & Sons, New York.

Stevenson, F.J., and E.T. Elliot. 1989. Methodologies for assessing the quantity and quality of soil organic matter. p. 173–199. *In* Dynamics of soil organic matter in tropical ecosystems. Univ. of Hawaii, Honolulu.

Stewart, J.W.B., and C.V. Cole. 1989. Influence of elemental interactions and pedogenic processes on soil organic matter dynamics. Plant Soil. 115:199–209.

Stewart, J.W.B., R.F. Follett, and C.V. Cole. 1987. Integration of organic matter and soil fertility concepts into management decisions. p. 1–8. *In* R.F. Follett et al. (ed.) Soil fertility and organic matter as critical components of production systems. SSSA Spec. Publ. 19. SSSA, Madison, WI.

Stewart, J.W.B., and A.N. Sharpley. 1987. Controls on dynamics of soil and fertilizer phosphorus and sulfur. p. 101–121. *In* R.F. Follett et al. (ed.) Soil fertility and organic matter as critical components of production systems. SSSA Spec. Publ. 19. SSSA, Madison, WI.

Stewart, J.W.B., and H. Tiessen. 1987. Dynamics of soil organic phosphorus. Biogeochemistry 4:41–60.

Tarafdar, J.C., and N. Classen. 1988. Organic compounds as a phosphorus source for higher plants through the activity of phosphatases produced by plant roots and microorganisms. Biol. Fertil. Soils 5:308–312.

Tarafdar, J.C., and A. Jungk. 1987. Phosphatase activity in the rhizosphere and its relation to the depletion of soil organic phosphorus. Biol. Fertil. Soils. 3:199–204.

Tate, K.R. 1985. Soil phosphorus. p. 329–377. *In* Soil organic matter and biological activity. Nijhof/Junk, Dordrecht.

Tate, K.R., and R.H. Newman. 1982. Phosphorus fractions of a climosequence of soils in New Zealand tussock grassland. Soil Biol. Biochem. 14:191–196.

Tate, K.R., D.J. Ross, A.J. Ramsay, and K.N. Whale. 1991. Microbial biomass and bacteria in two pasture soils: An assessment of measurement procedures, temporal variations, and the influence of P fertility status. Plant Soil. 132:233–241.

Tiessen, H. 1991. Characterization of soil phosphorus and its availability in different ecosystems. Trends Soil Sci. 1:83–99.

Tiessen, H., E. Frossard, A.R. Mermut, and A.L. Nyamekye. 1991. Phosphorus sorption and properties of ferruginous nodules from semi-arid soils from Ghana and Brazil. Geoderma. 48:373–389.

Tiessen, H., and J.O. Moir. 1983. Characterization of available P by sequential extraction. p. 75–86. *In* M.R. Carter (ed.) Soil sampling and methods of analysis. Lewis Publ., Chelsea, MI.

Tiessen, H., I.H. Salcedo, and E.V.S.B. Sampaio. 1992. Nutrient and soil organic matter dynamics under shifting cultivation in semi-arid northeastern Brazil. Agric. Ecosyst. Environ. 38:139–151.

Tiessen, H., and M.C.D. Santos. 1989. Variability of C, N and P content of a tropical semi-arid soil as affected by soil genesis, erosion and land clearing. Plant Soil. 119:337–341.

Tiessen, H., and J.W.B. Stewart. 1983. Particle size fractions and their uses in studies of soil organic matter: II. Cultivation effects on organic matter composition of size fractions. Soil Sci. Soc. Am. J. 47:509–514.

Tiessen, H., J.W.B. Stewart, and C.V. Cole. 1984. Pathways of phosphorus transformations in soils of differing pedogenesis. Soil Sci. Soc. Am. J. 48:853–858.

Tiessen, H., J.W.B. Stewart, and J.O. Moir. 1982. Cultivation effects on the amounts and concentration of carbon, nitrogen and phosphorus in grassland soils. Agron. J. 74:831–835.

Tiessen, H., J.W.B. Stewart, and J.O. Moir. 1983. Changes in organic and inorganic phosphorus composition of two grassland soils and their particle size fractions during 60–90 years of cultivation. J. Soil Sci. 34:815–823.

Van Veen, J.A., and P.J. Kuikman. 1990. Soil structural aspects of decomposition of organic matter by micro-organisms. Biogeochemistry. 11:213–274.

Van Veen, J.A., J.N. Ladd, and M.J. Frissel. 1984. Modelling C and N turnover through the microbial biomass in soil. Plant Soil. 76:257–274.

Walbridge, M.R., and P.M. Vitousek. 1987. Phosphorus mineralization potentials in acid organic soils: Processes affecting ^{32}-P isotope dilution measurements. Soil Biol. Biochem. 19:9–17.

Wilson, M.A. 1987. NMR techniques and applications in geochemistry and soil chemistry. Pergamon Press, Oxford.

Witter, E., A.M. Maternsson, and F.V. Garcia. 1993. Size of the soil microbial biomass in a long-term field experiment as affected by different N-fertilizers and organic manures. Soil Biol. Biochem. 25:659–669.

Woomer, P. 1992. Use of the CENTURY model to simulate phosphorus dynamics in tropical ecosystems. p. 232–239. *In* H. Tiessen and E. Frossard (ed.) Phosphorus cycles in terrestrial and aquatic ecosystems. Regional Workshop 4, Nairobi, Kenya. 18–22 Mar. 1991. Saskatchewan Inst. of Pedology, Univ. of Saskatchewan, Saskatoon.

9 Site-Specific Soil Tests and Interpretations for Potassium

D. J. Eckert
Ohio State University
Columbus, Ohio

Soil testing and test interpretation for K generally involve removing some fraction of soil K using one of several extractants and developing a fertilizer recommendation based on the quantity of K removed by that extractant. Several extractants are currently used in the USA, including 1 M NH$_4$OAc adjusted to pH 7.0 (Bray, 1944) and the Mehlich-3 combination extractant (Mehlich, 1984). These extractants remove soil K freely soluble in the soil solution and that held on cation exchange sites, a fraction often labeled exchangeable K. The quantity of K held in other forms and removed by these extractants is open to speculation.

THE GENERALIZED SUFFICIENCY LEVEL APPROACH

In the early 1940s, Bray (1944) noted that crop yield was related to the quantity of K removed from the soil by NH$_4$OAc and proposed modifying the Mitscherlich equation to describe the relationship between these variables. He also suggested that fertilizer K could be included as an additional term in the modified equation:

$$\text{Log}(A - y) = \text{Log } A - (c_1 x_1 + c_2 x_2) \quad [1]$$

y = yield at specific x_1 and x_2

A = maximum yield

x_1 = soil K concentration

x_2 = K added as fertilizer

c_1, c_2 = constants

This equation describes a response surface, the form of which is influenced by the constants c_1 and c_2. The values of these constants are determined by a

---

Copyright © 1994 Soil Science Society of America, 677 S. Segoe Rd., Madison, WI 53711, USA.
Soil Testing: Prospects for Improving Nutrient Recommendations, SSSA Special Publication 40.

number of factors, including the specific crop in question, the characteristics of the site on which the response is evaluated, and the weather during the season of evaluation. If it accurately reflects future conditions, an equation of this type allows one to calculate: (i) the exchangeable-K concentration that produces optimum yield without the addition of K fertilizer, the sufficiency level; and (ii) the quantity of K fertilizer needed to achieve optimum yield if the exchangeable-K concentration is below the sufficiency level.

This sufficiency level approach to soil testing and fertilizer recommendation for K became extremely popular and, in the last decade, has been the philosophy used most often as a basis for fertilizer recommendations at Land Grant Universities (Eckert, 1987). It has led to several general types of K fertilizer recommendations. One, a crop response recommendation, provides only enough additional K to produce a desired yield, and is made without regard to its effect on soil K concentrations after harvest. Such recommendations are often popular with those interested in minimizing fertilizer use, particularly in the short term. Much more common, however, are maintenance, build-up, and drawdown recommendations. Fertilizer programs based on maintenance should hold soil test K concentrations constant during cropping by replacing soil K removed by the crop. Those based on buildup raise K concentrations into a desired range over time by adding somewhat more K to the soil than is removed by the crop. Drawdown recommendations at less than the maintenance rate (or recommendations for no fertilizer) are often made when soil K concentrations are above the sufficiency level, the purpose being to reduce K fertility to more reasonable levels, while allowing the crop to feed on K already in the soil. Soil fertility specialists making any of these types of recommendations usually expect a K soil test to provide two needed pieces of information: (i) an indication of whether or not a crop is likely to respond to additions of K fertilizer; and (ii) how the soil K status changes in response to fertilizer addition.

The success of today's methods and interpretations in achieving the first objective is believed questionable in some situations. Since the time Bray popularized the concept of using exchangeable K and sufficiency to predict the K status of soils, several studies have indicated apparent failures of exchangeable-K measurements to predict soil-K availability or likelihood of response to K fertilizer (Chandler et al., 1945; Gholston & Hoover, 1948; Pope & Cheney, 1957; Skogley & Haby, 1981). These continuing observations, plus a general desire to improve soil testing methods and interpretations, have led to several alternative proposals for conducting and interpreting tests for available K in soils.

OTHER FACTORS INFLUENCING FERTILIZER RECOMMENDATIONS

Currently, predicting a response to fertilizer addition usually involves sampling a site, performing a soil test (usually an extraction), and interpreting the results of that test in terms of the research data most applicable to the situation at hand. Adjustments for specific conditions are made when possible. Few would argue that all crops should respond identically to the same K addition at the same soil test K concentration; thus, different recommendations for different crops are common. Additionally, other factors should also be considered when predicting responses, including site characteristics and cultural practices. (Determining the likelihood of *profitable* responses also involves additional cost-accounting considerations, which will not be addressed here.) Though weather during the growing season can affect fertilizer response from year to year, it is usually not considered in making recommendations because it cannot be predicted very far in advance. Site and cultural practice variables can be specified in advance; however, despite this ability, these variables have not always been considered when recommendations are made.

EXCHANGABLE POTASSIUM INTERPRETATIONS FOR SPECIFIC SOILS

Perhaps the most popular alternative to the sufficiency level approach has been to interpret exchangeable-K concentrations in terms of percentage base saturation or cation saturation ratios (Bear et al., 1945; Graham, 1959). This concept proposes that soils contain sufficient K when exchangeable-K concentrations are 2 to 5% of the cation-exchange capacity, an interpretation that allows some consideration of individual soil characteristics in the recommendation process. The concept and work investigating it have been reviewed previously (Eckert, 1987; McLean, 1977), and reviews show no real evidence that this method of interpretation offers any improvement over the sufficiency level approach.

A somewhat similar approach to improving predictions of K-supplying power has been to classify soils by properties that might influence K availability. Németh et al. (1970) demonstrated that, in a set of soils from northern Germany, concentrations of K in the soil solution (the K assumed available for immediate plant uptake) could be better related to exchangeable K if soils were grouped according to silt and clay concentrations. Fisher (1974) proposed that as the cation-exchange capacity of a soil increases (as a function of clay content), the quantity of exchangeable K needed for sufficiency should also increase:

$$SL_K = 110 + 2.5 \times CEC \qquad [2]$$

SL_K = K sufficiency level ($\mu g\ g^{-1}$)

CEC = cation-exchange capacity (cmol$_c$ kg^{-1})

This relationship was adopted for use (and is still used) in Ohio (McLean, 1976), with Fisher's original relationship being used for grass crops and slightly greater values adopted for legumes.

Not much later, McLean et al. (1979, 1982) proposed that further improvements in K testing and recommendations could be made by evaluating the K fixation-release characteristics of soils during the testing process. This proposal involved moist-equilibrating individual soil samples with a K solution and measuring the quantity of added solution K fixed by the soil against a subsequent

NH$_4$OAc extraction. A resulting index for that sample could then be used to calculate the quantity of K to be added to that individual soil to raise exchangeable K to the sufficiency level, with consideration of K fixation.

Such 1-yr buildup programs are rarely recommended in Ohio. The most common recommendations for K fertilization (Ohio Cooperative Extension Service, 1988) are based on field response trials that indicate the quantity of additional K needed to produce maximum yield at various exchangeable K concentrations on different soils. These quantities are then modified to ensure maintenance of soil test K concentrations at levels near sufficiency, effect some build-up when K concentrations are less than sufficiency, and allow drawdown when they are above sufficiency. Recommended fertilizer rates effect buildup and drawdown during a period of several years. Equations relating experimental yields to added K at different exchangeable K concentrations below sufficiency indicate that adding enough K to reach the sufficiency level is not usually needed to achieve maximum yield (Eckert & Johnson, 1991).

Comparing recommendations generated by McLean's method (including an added maintenance component) to current Ohio State University recommendations and those resulting strictly from crop response data shows that McLean's 1-yr build-up program requires more K fertilizer than the others in low fertility situations (Table 9–1). Difficulty justifying these high rates from a short-term crop-response standpoint will probably limit McLean's approach as a basis for making recommendations; however, the method, itself, could prove valuable as an analytical tool. A major benefit of the method may be its allowing laboratories to estimate K fixation tendencies for individual soils rather quickly and easily, providing some answer to the question of how soil test levels on individual soils might change upon fertilization.

Despite their seeming differences, all of these proposed improvements over Bray's original approach have relied on the assumption that exchangeable K (and often only that in the upper 20 cm of the soil) provides an acceptable and useful measure of the K supplying power of a site under a wide range of conditions. This assumption has worked well in some instances, but has not proven acceptable in all situations. In some cases, special site- or soil-specific interpretations can

Table 9–1. K fertilizer rates recommmended to produce a 9.5 Mg ha^{-1} corn yield on a soil with a cation-exchange capacity of 10 cmol$_c$ kg^{-1}.

Exchangeable K	Crop response[†]	OSU[‡]	McLean[§]
µg g^{-1}		kg ha^{-1}	
50	40	100	230
75	0	80	160
125	0	50	40
175	0	40[¶]	40
225	0	0	0

[†] K required only to achieve yield (Eckert and Johnson, 1991).
[‡] Ohio State University recommedation, including maintenance, buildup, or drawdown (Ohio Corporative Extension Service, 1988).
[§] From McLean (1982) plus maintenance.
[¶] Maintenance (crop removal) rate.

overcome some of the difficulty (for example, considering the K supplying power of subsoil not taken during sampling). In other cases, alternative interpretations, including Beckett's quantity-intensity system (Q/I; Beckett, 1964), and alternative strategies for removing K from soils, including electroultrafiltration (EUF) (Németh, 1979; Mengel & Uhlenbecker, 1993), or boiling 1 M HNO_3 (Pratt & Morse, 1954) have been suggested as means of improving K soil tests. In the USA none of these approaches has shown enough potential for improving recommendations to attract the serious attention of the commercial soil testing industry, though EUF has gained some popularity in parts of Europe.

At present many soil testing laboratories are adopting the Mehlich-3 extractant in lieu of the traditional NH_4OAc. The objective of this change has been to adopt an extractant that allows simultaneous extraction and determination of several elements in one procedure. Data from Pennsylvania (Beegle & Oravec, 1990) and Ohio (D.J. Eckert & M.L. Watson, 1988, unpublished data) show that these extractants remove virtually the same quantity of K from a given soil sample (Ohio data shown in Fig. 9–1). Therefore, despite potential cost savings and improvements in laboratory efficiency, merely switching from NH_4OAc to Mehlich-3 seems unlikely to improve predictability of fertilizer responses.

Extraction-based strategies might be criticized on grounds that the extraction process is not representative of the true relationship between soils and plants, and does not reflect differences in this relationship between sites. Plant roots do not strip an entire single fraction of K, in situ, from the soil during the growing season, nor do K pools in the soil remain static during K removal by roots. With this in mind, Corey (1987) has suggested that a more mechanistic approach be taken to soil testing, using newer models of soil-plant relationships. Such models are available (Baldwin et al., 1973; Barber, 1984; Barber & Cushman, 1981; Nye & Tinker, 1977), and have shown some promise in predicting actual K uptake by corn (*Zea mays* L.) in the laboratory (Claassen & Barber, 1976; Shaw et al., 1983)

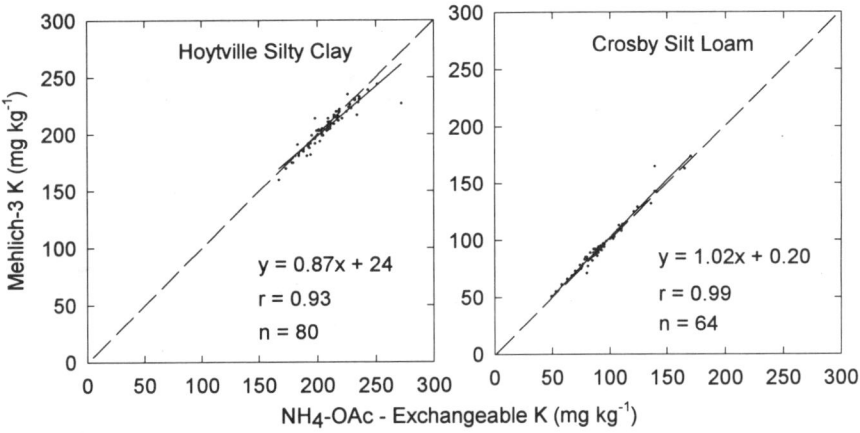

Fig. 9–1. Potassium extracted from two Ohio soils by NH_4OAc and the Mehlich-III extractant.

and by soybean [*Glycine max* (L.) Merr.] in the field (Silberbush & Barber, 1984). Given our increasing understanding of how the soil actually supplies K to plants, it is perhaps appropriate that the concepts inherent in such models be considered as foundations for alternative approaches to predicting K sufficiency and fertilizer needs. The Cushman-Barber model, for example, uses three soil characteristics to predict nutrient uptake. These are the initial concentration of the nutrient in the soil solution, the ability of the soil to replenish the solution as the nutrient is depleted, and the rate at which the nutrient diffuses to the root. All of these factors may vary considerably from site to site, but none is determined directly by the type of extractions used in routine soil testing.

A newer and more mechanistic approach, developed by Skogley et al. (1990), uses a mixed-bed ion-exchange-resin sink to remove K from soils. These researchers have proposed that the quantity of K removed from a saturated soil paste by this sink (usually resin beads encased in a mesh ball) provides an indication of soil solution K concentration and rate of K diffusion through the soil, and provides a better estimate of K-supplying power than a simple NH_4OAc extraction. Some preliminary work has shown that the quantity of K extracted by this method was related to K uptake in a limited number of trials; however, more investigation, particularly in the field, will be required before the method can be judged applicable to a wide variety of situations.

THE CHALLENGES OF CONSERVATION TILLAGE

The entire approach of using simple, in-laboratory soil tests as stand-alone methods for determining fertilizer needs, however, may be complicated by the adoption of conservation tillage practices, particularly no-tillage production, by growing numbers of farmers. Such practices can accentuate the differences between the whole field soil in which plants grow and the soil material sent to the laboratory for evaluation. Under conservation tillage conditions, K may become stratified in the upper levels of the soil profile (Table 9–2), with exchangeable K concentrations decreasing as depth increases (Eckert, 1985, 1991). This uneven distribution of K may confound test interpretations based on the assumption of uniform distribution of K throughout the depth of sampling. For example, in situations where K concentrates very close to the soil surface, moisture relations near the surface may prove quite important in affecting K uptake. Yibirin et al. (1993)

Table 9–2. Distribution of exchangable K in two Ohio soils after 4 yr of no-till cropping.

Depth	Soil	
	Canfield silt-loam†	Hoytville silt, clay‡
cm	———— mmol kg^{-1} ————	
0–5	4.06	88.8
5–10	2.03	5.19
10–15	1.85	4.69
15–20	1.73	4.68

† Fine-loamy, mixed, mesic Aquic Fragiudalf.
‡ Fine, illitic, mesic Mollic Ochraqualf.

have demonstrated that K uptake by no-till corn grown on a crusting ochraqualf without surface residue cover was less than that grown with residue, an effect presumed due to the moisture conserving effect of the mulch. Such findings show that adoption of conservation tillage may force a need to characterize more site- and season-specific factors to maintain and/or improve the reliability of K recommendations.

AN INTEGRATED APPROACH

Ultimately, K fertilizer recommendations may be made based on the results of a number of evaluations (Fig. 9–2). Some of these evaluations, including crop to be grown and immediate soil conditions (residue cover, surface infiltration capacity, or degree of compaction) might be estimated yearly, by site inspection. Other information such as the K supplying power and fixation capacity of the soil material located at the site might be determined in the laboratory. Such evaluations might involve determination of values needed to apply mechanistic nutrient uptake models, or other factors related to them. These data might then be combined with data from existing or new databases (soil surveys or weather records) and processed by computer models that would generate a fertilizer recommendation derived almost entirely from site-specific data. Hopefully, these recommendations would allow farmers to make more profitable decisions, and would provide continuing credibility for the soil-testing industry.

While adoption of such an approach might yield long-range benefits to production agriculture, its development would require major shifts in the approach most have traditionally taken to soil testing and nutrient recommendation. The traditional approach has been to develop extractants and use the results of extraction, with perhaps one or two additional pieces of information, to generate recommendations. The trend has been toward simplicity and developing least-cost procedures. While a more comprehensive and mechanistic approach may well produce more accurate recommendations, it will probably be more expensive to

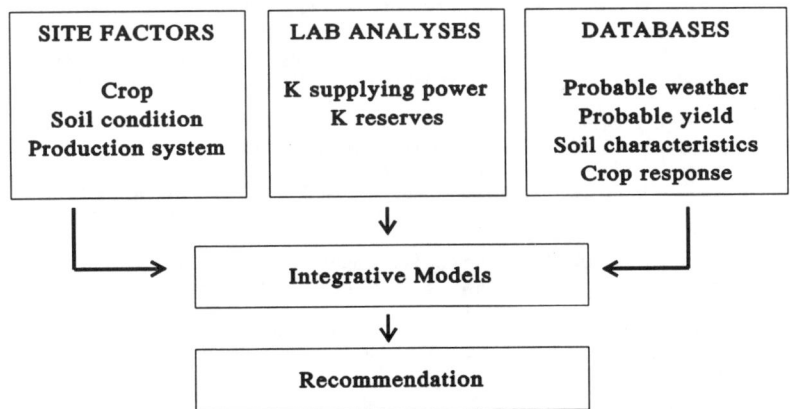

Fig. 9–2. An integrative model for generating improved site-specific recommendations for K fertilizer.

the user. Whether such a recommendation scheme is ever developed and used will depend largely on whether users perceive that they derive any greater benefit from it than from existing schemes. Such benefits will probably be determined by profitability.

REFERENCES

Baldwin, J.P., P.H. Nye, and P.B. Tinker. 1973. Uptake of solutes by multiple root systems from soil: III. A model for calculating the solute uptake by a randomly dispersed root system developing in a finite volume of soil. Plant Soil 38:621–635.

Barber, S.A. 1984. Soil nutrient bioavailability. Wiley-Interscience, New York.

Barber, S.A., and J.H. Cushman. 1981. Nitrogen uptake model for agronomic crops. p. 382–409. *In* I.K. Iskandar (ed.) Modeling waste water—land treatment. Wiley-Interscience, New York.

Bear, F.E., A.L. Prince, and J.L. Malcolm. 1945. The potassium needs of New Jersey soils. New Jersey Agric. Exp. Stn. Bull. 721. New Brunswick.

Beckett, P.H.T. 1964. Studies on soil potassium: II. The immediate Q/I relations of labile potassium in the soil. J. Soil Sci. 15:9–23.

Beegle, D.B., and T.C. Oravec. 1990. Comparison of field calibrations for Mehlich 3 P and K with Bray-Kurtz P1 and ammonium acetate K for corn. Commun. Soil Sci. Plant Anal. 21:1025–1036.

Bray, R.H. 1944. Soil plant relations: I. The quantitative relation of exchangeable potassium to crop yields and to crop response to potash additions. Soil Sci. 58:305–324.

Chandler, R.F., M. Peech, and C.W. Chang. 1945. The release of exchangeable and nonexchangeable potassium from different soils upon cropping. Agron. J. 37:709–721.

Claassen, N., and S.A. Barber. 1976. Simulation model for nutrient uptake from soil by a growing plant root system. Agron. J. 68:961–964.

Corey, R.B. 1987. Soil test procedures: Correlation. p. 15–22. *In* J.R. Brown (ed.) Soil testing: Sampling, correlation, calibration, and interpretation. SSSA Spec. Publ. 21. SSSA, Madison, WI.

Eckert, D.J. 1985. Review: Effects of reduced tillage on the distribution of soil pH and nutrients in soil profiles. J. Fert. Issues 2:86–90.

Eckert, D.J. 1987. Soil test interpretations: Basic cation saturation ratios and sufficiency levels. p. 53–64. *In* J.R. Brown (ed.) Soil testing: Sampling, correlation, calibration, and interpretation. SSSA Spec. Publ. 21. SSSA, Madison, WI.

Eckert, D.J. 1991. Chemical attributes of soils subjected to no-till cropping with rye cover crops. Soil Sci. Soc. Am. J. 55:405–409.

Eckert, D.J., and J.W. Johnson. 1991. Fertility management. p. 19–32. *In* D.J. Eckert (ed.) Crop production alternatives. Ohio Coop. Ext. Serv. Bull. 812. Ohio Coop. Ext. Serv., Columbus, OH.

Fisher, T.A. 1974. Some considerations for interpretations of soil tests for phosphorus and potassium. Missouri Agric. Exp. Stn. Res. Bull. 1007. Columbia.

Graham, E.R. 1959. An explanation of theory and methods of soil testing. Missouri Agric. Exp. Stn. Bull. 734 Columbia.

Gholston, L.E., and C.D. Hoover. 1948. The release of exchangeable and non-exchangeable potassium from several Mississippi and Alabama soils upon continuous cropping. Soil Sci. Soc. Am. Proc. 13:116–121.

McLean, E.O. 1976. Exchangeable K levels for maximum crop yields on soils of different cation exchange capacities. Commun. Soil Sci. Plant Anal. 7:823–838.

McLean, E.O. 1977. Contrasting concepts in soil test interpretation: Sufficiency levels of available nutrients versus basic cation saturation ratios. p. 39–54. *In* T.R. Peck et al. (ed.) Soil testing: Correlating and interpreting the analytical results. ASA Spec. Publ. 29. ASA, CSSA, and SSSA, Madison, WI.

McLean, E.O., J.L. Adams, and R.C. Hartwig. 1982. Improved corrective fertilizer recommendations based on a two-step alternative usage of soil tests: II. Recovery of soil-equilibrated potassium. Soil Sci. Soc. Am. J. 46:1198–1201.

McLean, E.O., T.O. Oloya, and J.L. Adams. 1979. Soil tests to inventory the initially available levels and to assess the fates of added P and K as bases for improved fertilizer recommendations. Commun. Soil Sci. Plant Anal. 10:623–630.

Mehlich, A. 1984. Mehlich 3 soil test extractant: A modification of Mehlich 2 extractant. Commun. Soil Sci. Plant Anal. 15:1409–1416.

Mengel, K., and K. Uhlenbecker. 1993. Determination of available interlayer potassium and its uptake by ryegrass. Soil Sci. Soc. Am. J. 57:761–766.

Németh, K. 1979. The availability of nutrients in the soil as determined by electro-ultrafiltration (EUF). Adv. Agron. 31:155–188.

Németh, K., K. Mengel, and H. Grimme. 1970. The concentration of K, Ca and Mg in the saturation extract in relation to exchangeable K, Ca and Mg. Soil Sci. 109:179–185.

Nye, P.H., and P.B. Tinker. 1977. Solute movement in the soil-root system. Univ. of California Press, Berkeley.

Ohio Cooperative Extension Service. 1988. Ohio agronomy guide. 12th ed. Ohio Coop. Ext. Serv., Columbus, OH.

Pope, A., and H.B. Cheney. 1957. The potassium supplying power of several western Oregon soils. Soil Sci. Soc. Am. Proc. 21:75–79.

Pratt, P.F., and H.H. Morse. 1954. Potassium release from exchangeable and nonexchangeable forms in Ohio soils. Ohio Agric. Exp. Stn. Res. Bull. 747. Wooster.

Shaw, J.K., R.K. Stivers, and S.A. Barber. 1983. Evaluation of differences in potassium availability in soils of the same exchangeable potassium level. Commun. Soil Sci. Plant Anal. 14:1035–1049.

Silberbush, M., and S.A. Barber. 1984. Phosphorus and potassium uptake of field-grown soybean cultivars predicted by a simulation model. Soil Sci. Soc. Am. J. 48:592–595.

Skogley, E.O., S.J. Georgitis, J.E. Yang, and B.E. Schaff. 1990. The phytoavailability soil test—PST. Commun. Soil Sci. Plant Anal. 21:1229–1243.

Skogley, E.O., and V.A. Haby. 1981. Predicting crop responses on high-potassium soils of frigid temperature and ustic moisture regimes. Soil Sci. Soc. Am. J. 45:533–536.

Yibirin, H., J.W. Johnson, and D.J. Eckert. 1993. No-till corn production as affected by mulch, potassium placement, and soil exchangeable potassium. Agron. J. 85:639–644.

10 Effects of Iron Oxidation State on the Fate and Behavior of Potassium in Soils

Siyuan Shen and Joseph W. Stucki
University of Illinois
Urbana, Illinois

Potassium is one of the most important plant nutrients in soils, and has thus been studied extensively (Potash & Phosphate Institute, 1980; Munson, 1985). Despite much study and diligent efforts, the fundamental chemical and physical phenomena that govern its fate, movement, and availability to plants in soils have yet to be characterized fully (Bertsch & Thomas, 1985). Soil tests for K often fail to reveal the true fertilizer demand in the field, resulting in frequently unreliable and inefficient fertilizer recommendations (Munson, 1980; T.R. Peck, Univ. of Illinois Soil Testing Laboratory, 1992, personal communication). Perhaps the major factor making the solution to this problem elusive is that soil K is usually distributed among soluble, exchangeable, fixed, and insoluble forms, and becomes redistributed among these forms in an unpredictable manner. In other words, the fate and movement of K in the soil have yet to be well understood. Because soil tests measure primarily only the soluble and exchangeable forms of K, any uncontrolled or unrecognized transitions of K among its forms between the time of testing and the time of attempted uptake by the plant roots will produce an erroneous prediction of available K.

Previous studies provided empirical evidence that K behavior is correlated to a number of different soil and environmental factors (Munson, 1980). These include the types of soil minerals present (Rich, 1964), moisture regime (Barber, 1960; Bates & Scott, 1964; Mengel & von Branschweig, 1972), cropping and fertilizer history (Doll et al., 1965; T.R. Peck, 1992, personal communication; Kaspar et al., 1989), temperature fluctuations (Sparks & Liebhardt, 1982), and weathering (Mortland et al., 1956). No unified explanation linking all of these variables in a consistent manner has been established.

Soil testing methods have evolved over the years. As science recognized little relationship between crop available and total K in the soil (Reitemeier et al., 1951), the exchangeable form of K was used to estimate plant-available forms. But, again, response predictions were wrong about as often as they were right. Hence, an improved soil test method was proposed by McLean et al. (1982) using the quantity–intensity (Q/I) concept advanced by Beckett (1964). This test is based on the ratio of total K (Q) to an intensity (I) term, which is the quotient of

---

Copyright © 1994 Soil Science Society of America, 677 S. Segoe Rd., Madison, WI 53711, USA.
Soil Testing: Prospects for Improving Nutrient Recommendations, SSSA Special Publication 40.

thermodynamic activities of K, Ca, and Mg. While this method is somewhat more rigorous from a chemical and mathematical viewpoint than other methods, it is still largely empirical insofar as the mechanism for K availability is concerned. The concept recognizes the fact that the soil matrix consists of a heterogeneous array of energy sites for K retention, and those sites that bind with lower energies will more readily release K to the plant. Using this principle of varying binding energies, electroultrafiltration (EUF) has been applied to differentiate the chemical dynamics of soil K in terms of nutrient availability (Grimme & Nemeth, 1979). Sparks and Jardine (1981) investigated the kinetics of K adsorption and desorption in a soil, and provided further rigorous chemical methods (Sparks & Huang, 1985) for treating K data. But again, the mechanisms are incomplete.

One puzzling observation is that the amount of exchangeable K in a routine soil test changes with moisture content—sometimes increasing (Bohannon, 1957) and sometimes decreasing (Bates & Scott, 1964). One explanation that is sometimes suggested for increased exchangeability upon drying is that K^+ is expelled from the interlaminar spaces of the clay minerals along with the evaporating water; but this appears to contradict fundamental electrostatic neutrality requirements. Bates and Scott (1964) found that low-volatile organics and sucrose both inhibit release of K^+, which they explained as the result of diffusion inhibition.

These and other results indicate that other processes may be involved which have yet to receive adequate consideration. One such process is oxidation and reduction of expanding soil minerals. Oxidation and reduction (redox) is the terminology used to describe the chemical process that changes the electrical charge of a chemical element. Some of the elements in soils that are susceptible to redox changes are Fe, Mn, N, and S, and such processes are commonly found in agricultural soils as a result of biological activity and alternate wetting and drying events. The diffusion inhibition mechanism proposed by Bates and Scott (1964) may actually be the result of microbial growth and reducing conditions in the soil, producing a change in redox status.

Redox reactions involving the above-mentioned elements, and particularly Fe, are responsible for many changes in the physical and chemical properties of soils (Yu, 1985). While the transformation of N to its various forms, ranging from NO_3^- to NH_4^+, is probably the most-studied redox process in soils, significant changes in oxidation state also occur in many soil minerals because of the presence of Fe in their crystal structures. The weathering of primary minerals is an example of an oxidation process in which Fe^{2+} is converted to more expansive types which release various nutrients, including K^+, to the soil solution. The in situ reduction of Fe^{3+} to Fe^{2+} in secondary minerals (vermiculite, montmorillonite, and illite) also occurs, and creates a climate in which some of the beneficial effects of weathering may be reversed, such as the fixation of K^+. This is due to numerous changes that are invoked in the physicochemical properties of the mineral phase of soils, including swellability in water (Stucki et al., 1984), electrical charge (Stucki & Roth, 1977; Lear & Stucki, 1985), and surface area (Lear & Stucki, 1989). These redox phenomena are of great importance to soil fertility because the availability of plant nutrients depends in large degree on the surface chemistry of the minerals. Since the oxidation state will vary with different environmental conditions, the associated properties also may change significantly throughout the year. The purpose of this chapter is to review the present state of knowledge of the role of Fe oxidation state in determining K availability in soils.

LEVELS OF IRON REDUCTION IN CLAYS

The levels of Fe^{2+} attained by the chemical reduction of six clay samples are reported in Table 10–1 (Shen & Stucki, 1994, unpublished data). Sample SWa-1 contained the most total Fe of all samples, and the Drummer soil (fine-silty, mixed, mesic Typic Haplaquoll) from Urbana was highest among the soil clays. The Fe_{2+} content of these samples in their initial, oxidized (unaltered) state varied from <0.16% of the total Fe in smectite to >25% in Fithian illite. The initial Fe^{2+} content of the soil clays was similar to smectite API 25, ranging from 3% to 10% of their total Fe.

When Shen and Stucki (1994, unpublished data) reduced samples using sodium dithionite, the Fe^{2+} content exceeded 60% of total Fe in all samples, except the Drummer soil from Dekalb, IL, contained <25%, which they attributed to some reoxidation during the post-reduction washing procedure. The levels of Fe^{2+} achieved were not necessarily the limiting case because longer treatments are known to reduce most of the Fe in sample SWa-1 (Komadel et al., 1990), which was only reduced 74% by the treatment reported in Table 10–1.

CATION FIXATION BY PHYLLOSILICATES

Smectites

Recent laboratory studies of clay suspensions in electrolyte solutions showed that the change in surface charge that occurs during Fe reduction increases the ability of the mineral to fix interlayer cations (Fig. 10–1), including Na^+ (Lear &

Table 10–1. Iron contents of oxidized, reduced, and reoxidized reference clays and Illinois soil clays.

		Fe^{2+}		Total Fe	
Treatment	Sample†	mmol g^{-1}	% of total Fe	mmol g^{-1}	% of sample
Oxidized	SWa-1	0.005	0.16	3.166	17.68
	API 25	0.026	5.78	0.456	2.55
	Fithian	0.211	25.4	0.830	4.64
	Drummer (D)	0.078	9.50	0.824	4.61
	Drummer (U)	0.076	6.94	1.100	6.14
	Cisne	0.033	3.15	1.035	5.78
Reduced	SWa-1	2.386	74.0	3.141	17.54
	Fithian	0.464	60.8	0.764	4.27
	Drummer (D)	0.156	20.1	0.777	5.90
	Drummer (U)	0.709	66.8	1.054	5.90
	Cisne	0.614	61.5	0.999	5.58

†D = soil from DeKalb, IL; U = soil from Urbana, IL.

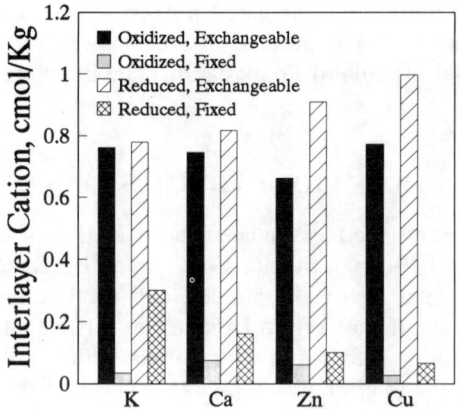

Fig. 10–1. Effect of Fe oxidation state on cation fixation (from Khaled & Stucki, 1991).

Stucki, 1989), Ca^{2+}, Cu^{2+}, and Zn^{2+} (Khaled & Stucki, 1991), and K^+ (Chen et al., 1987). Chen et al. (1987) observed a rapid increase in the nonexchangeable form of K^+ with Fe^{2+} content of freeze-dried soil clays and standard reference clays, wherein the total fixation capacity reached as high as 30% of the total exchange capacity. Khaled and Stucki (1991) followed the distribution of K^+ between exchangeable and nonexchangeable fractions with increasing Fe^{2+} content of Upton montmorillonite gels that were never dried, and found that fixation increased steadily with increasing Fe^{2+}, but that the exchangeable fraction remained almost constant (Fig. 10–2).

Khaled and Stucki (1991) also offered the following sample calculation showing the potentially great effect of oxidation state on the availability of K in

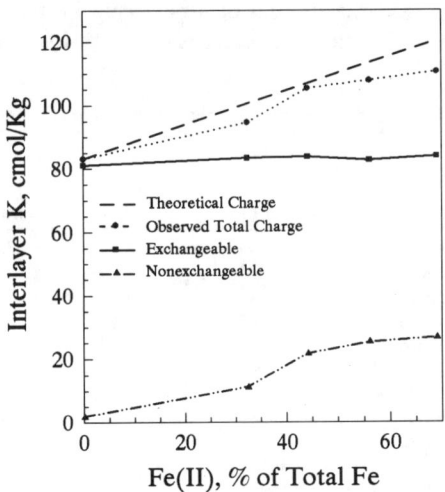

Fig. 10–2. Effect of Fe oxidation state on exchangeable, fixed, and total K in Upton montmorillonite (API 25) compared with total theoretical layer charge (from Khaled & Stucki, 1991).

soils. By interpolating the data in Fig. 10–2 to a reduced level of only 20% (% Fe(II) to total Fe = 20), the estimated amount of fixation is ≈10 meq 100 g^{-1} of clay. In a soil having a clay content of 15% (w/w), of which two-thirds is smectite, the K$^+$ fixation capacity would be ≈165 kg K$_2$O ha^{-1} 15 cm. Reduction levels may greatly exceed this value in the soil under various environmental conditions, so the potential for reduced clay to serve as a sink for K is tremendous. This point is even more poignant in view of the fact that a high application rate for K fertilizer is in the range of 37 kg K$_2$O ha^{-1}. A modest change in the oxidation state of Fe in the mineral could, therefore, remove virtually all of such an application from plant availability—at least temporarily.

Recent studies (Shen & Stucki, 1994, unpublished data) of two reference smectite clays (SWa-1 and API 25), one reference illite (Fithian), and the clay fractions from the A horizon of three Illinois soils [Drummer from Dekalb, Drummer from Urbana, and Cisne (fine, montmorillonitic, mesic Mollic Albaqualf) from Brownstown], all in electrolyte suspensions, confirmed that the total and fixed K contents of smectites increased when the clays were reduced, while the exchangeable K contents remained constant (Table 10–2). This result is consistent with observations by Chen et al. (1987) and Khaled and Stucki (1991). So we can generalize by stating that structural Fe reduction increases K fixation in smectites, and thus, probably decreases K availability in smectite soils.

Illite

The Fithian illite sample behaved very differently from the smectites. The amount of fixed K decreased upon treatment with the reducing agent, while K$_e$, the exchangeable K, increased (Table 10–2). The total K content also decreased. Soils containing a mixture of illite and smectite would, therefore, either fix or release K during reducing conditions depending on which clay dominates; or, if these minerals are precisely balanced, no change in K availability would occur because the amount that one releases would be fixed by the other. One possible explanation for the behavior of illite is that as more Fe^{2+} is added to the octahedral sheet during further reduction, the dipole moment of structural OH$^-$ groups become tilted more along the c-axis (the axis perpendicular to the clay plates) and thereby create a weaker attractive (or stronger repulsive) force between the clay layer and the interlayer K ions (Juo & White, 1969). Release of K to soil solution subsequently would be increased.

Table 10–2. Total, fixed, and exhangeable K contents of reference clays (n = 4).

Sample	Treatment	Fe^{2+} % of total Fe	Total K	Fixed K cmole Kg^{-1}	Exchangeable K
SWa-1	Unreduced	0.16	93.0	2.9	90.1
	Reduced	74.0	122.3	31.5	90.8
API 25	Unreduced	5.8	79.5	3.0	76.4
	Reduced	66.8	102.5	25.3	77.2
Fithian Illite	Unreduced	25.4	151.4	135.7	15.7
	Reduced	60.8	122.7	89.8	23.9

Soil Clays

Shen and Stucki (1994, unpublished data) also found (Table 10–3) that the total K content of both Drummer soils was constant during one redox cycle, but it varied slightly in the Cisne soil. The distribution of K between exchangeable and fixed fractions, however, fluctuated measurably, but less than in the pure clay samples. The behavior of these soil clays appeared to be closer to that of illite than of smectite, suggesting that the soil clays may be dominated by illite. X-ray powder diffraction (Fig. 10–3) revealed that the clay fraction of these three soils was about two-thirds illite and one-third smectite.

Iron oxides in the soil may significantly influence K availability. Shen and Stucki (1994, unpublished data) studied soils with and without Fe oxides present. The Fe oxides in the clay fraction of three soil samples were removed by citrate-bicarbonate-dithionite (CBD) treatment before the reduction experiment. Results from this experiment (Table 10–4) revealed that after oxides were removed from the soil clays, both exchangeable and fixed K increased when the clays were reduced, resulting in an increase in total K as well. Hence, removal of Fe oxides caused the soil clays to behave somewhat more like smectite than like

Table 10–3. Total, fixed, and exchangeable K contents of three Illinois soil clays (with oxides).

Soil clay	Treatment	Fe^{2+} % of total Fe	Total K	Fixed K	Exchangeable K
				cmole Kg^{-1}	
Drummer (DeKalb)	Unreduced	9.50	102.4	72.8	29.3
	Reduced	20.1	105.2	66.6	38.6
Drummer (Urbana)	Unreduced	6.94	105.3	71.6	33.7
	Reduced	66.8	105.3	65.7	39.6
Cisne (Brownstone)	Unreduced	3.15	78.84	43.97	30.9
	Reduced	61.5	82.1	39.2	42.9

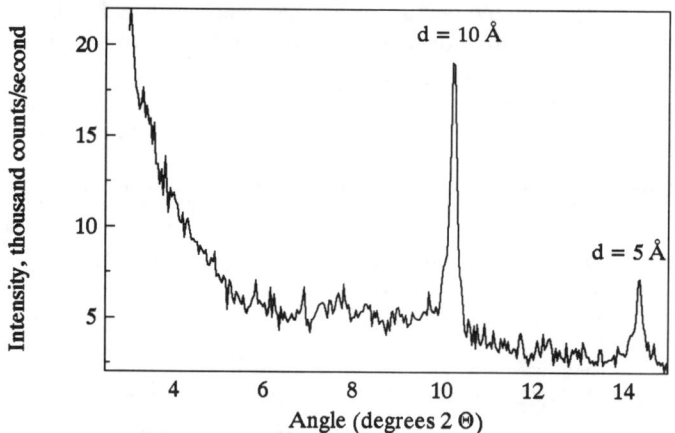

Fig. 10–3. X-ray powder diffraction pattern of <2-μm fraction of Cisne soil (Co K radiation).

Table 10–4. Total, fixed, and exchangeable K contents of three Illinois soil clays (oxides removed, $n = 4$).

Soil clay	Treatment	Fe^{2+} % of total Fe	Total K	Fixed K	Exchangeable K
				cmole Kg^{-1}	
Drummer (DeKalb)	Unreduced	1.58	119.3	86.1	33.2
	Reduced	19.2	125.2	88.5	35.7
Drummer (Urbana)	Unreduced	13.6	124.4	88.2	36.2
	Reduced	18.0	132.0	90.7	41.3
Cisne (Brownstone)	Unreduced	9.87	94.34	59.87	34.5
	Reduced	13.7	115.9	61.6	54.3

illite. Although the difference in fixed K was not very large between the reduced and nonreduced samples, it was significant for the levels of Fe reduction achieved in the samples. Chen et al. (1987) achieved similar amounts of fixed K when they reduced and freeze-dried the soil samples in their experiment.

Linear regression analysis reported in Table 10–5 reveals the quantitative relationship between fixed or exchangeable K and the Fe^{2+} content of the various soil clays. Notice that the correlation between fixed K and Fe oxidation state was excellent in the two Drummer soils, but was rather poor in the Cisne soil. We attribute the latter to variability in the Cisne sample.

EFFECTS OF REDOX CYCLES

All samples, including K-saturated reference and soil clays with oxides, were subjected to cyclic reduction and reoxidation treatments. The reoxidation treatment consisted of bubbling O_2 gas through the reduced suspension for 30 min, exposing it to air for 3 d, or adding sodium hypochlorite solution to the reduced suspension. The Fe^{2+} and fixed K contents in the reduced and reoxidized samples were measured with each cycle, and varied depending on the mineralogy.

Smectite

The level of Fe^{2+} in ferruginous smectite, where Fe ions are clustered within the clay crystal, is reflected in the intervalence electron transfer (IT) band observed at ~730 nm (Lear & Stucki, 1987). The continuous monitoring of the IT

Table 10–5. Linear regression parameters for the relation state of $K = A + B* Fe^{2+}$, where units of K are meq 100 g^{-1}, and of Fe^{2+} are mmol g^{-1}.

Soil	State of K	A	B	R^2
Drummer (DeKalb)	exchangable	21.0	0.773	0.899
	fixed	74.6	0.729	0.765
Drummer (Urbana)	exchangable	20.4	1.166	0.990
	fixed	81.2	0.518	0.849
Cisne (Brownstone)	fixed	53.6	0.601	0.337

band in Fe-rich smectite (SWa-1) suspensions revealed that this band increases in intensity with the number of Fe^{2+}–O–Fe^{3+} linkages in the clay structure, and reaches a maximum at ~45% of total Fe reduced to Fe^{2+} (Lear & Stucki, 1987; Komadel et al., 1990).

Spectrum A in Fig. 10–4 shows that in the first 10 to 20 min of reduction, ~45% of structural Fe was reduced in the K-saturated SWa-1 clay. But >3 h was required before an additional 20% of structural Fe was reduced. When O_2 was introduced into the suspension to reoxidize the clay, this latter 20% of Fe^{2+} was oxidized quickly (in ~0.5 h). But the 45% reduced initially was more difficult to reoxidize. Even after a 6-h exposure to O_2 (extension of results shown in Fig. 10–4, Spectrum A) ~20% of structural Fe was still in the Fe^{2+} state. Chemical analysis (method of Komadel et al., 1988) of the samples treated similarly revealed that 17% of the Fe remained as Fe^{2+} after 6 h of reoxidation. By contrast, the Na-saturated SWa-1 was completely reoxidized in 0.5 h (Komadel et al., 1990). The difference between Na- and K-saturated SWa-1 reduced in suspension indicates that K fixed in the interlayer of the clay during reduction can protect the structural Fe^{2+} from reoxidation to a large extent. This probably is due to the collapse of superimposed clay layers to the exclusion of interlayer water.

Six redox cycles were completed with Fe-rich smectite SWa-1. In Spectra B and C of Fig. 10–4 are reported the results from the third and fifth redox cycles of the same sample described by Spectrum A. The rate of initial reduction and the time required for completion of the reaction both decreased with increasing number of cycles, presumably because the Fe^{2+} contents were a little higher at the end of each cycle, and thus less Fe^{3+} remained in the sample. The first peak maximum decreased with each cycle, indicating that the available number of Fe^{2+}–O–Fe^{3+}

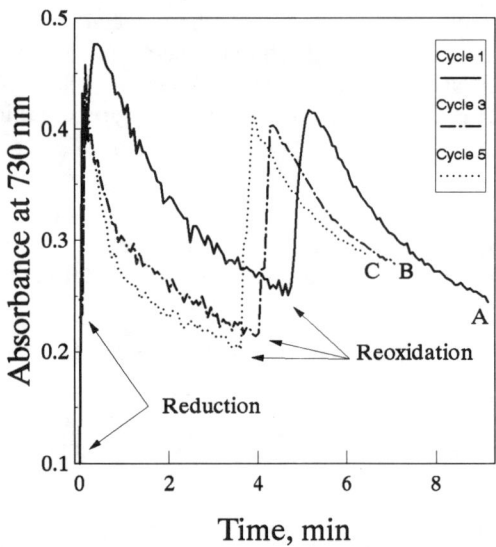

Fig. 10–4. Intensity of intervalence electron transfer transitions (IT) at 730 nm in ferruginous smectite (SWa-1) after multiple redox cycles.

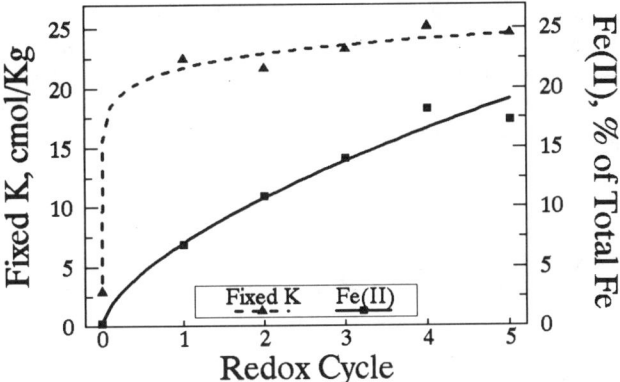

Fig. 10–5. Effect of redox cycles on Fe^{2+} and fixed K contents of ferruginous smectite (SWa-1).

pairs in the sample was less after each cycle. The Fe^{2+} remaining in the clay at the end of each cycle, therefore, must be clustered and uninvolved in subsequent redox cycles.

In Fig. 10–5 the residual Fe^{2+} that is present after the reoxidation step of the respective redox cycles, as determined by wet chemical analysis (Komadel et al., 1988), is plotted vs. redox cycle number. Notice that the residual Fe^{2+} level increases steadily to ~24% of the total Fe. The rate of Fe reduction appears to slow after five or six cycles, but conceivably the trend of increasing residual Fe^{2+} with each cycle could continue indefinitely to produce Fe^{2+} contents similar to that observed in Fithian illite (60%, Table 10–2).

Also reported in Fig. 10–5 is the amount of K that is nonexchangeable or fixed in the clay after several redox cycles. Reduction of the Fe increases fixed K in the smectite from ~2.5 to ~35 meq 100 g^{-1}. Reoxidation decreases the amount of fixed K to only ~25 meq 100 g^{-1}, thus failing to release all of the K that was fixed by the Fe reduction reaction. Subsequent redox cycles produced similar values upon reduction and reoxidation, indicating that a stable level of K fixation apparently is reached immediately with the first reduction treatment. Similar results were found for smectite API 25. Perhaps the combination of K fixation and the gradually increasing residual Fe^{2+} in the reoxidized clays could constitute a mechanism for illite formation in soils. In any case, these processes certainly will alter the kinetics of K availability in smectitic soils.

Illite and Soil Clays

Fithian illite changed color from dark gray to white and the clay fraction of Cisne soil changed color from a yellowish brown to light gray when reduced. No intervalence transfer (IT) band was observed in these clays because the Fe ions in the clay structure were not located in adjacent crystallographic sites, but charge transfer (CT) did occur between O and Fe^{3+}, giving a change in visible absorbance intensity at ~400 nm. The approximate position of the band for this electronic transition is 245 nm, and only the tail is seen at 400 nm. The peak itself cannot be studied by this system because the reducing agent absorbs ultraviolet light at approximately the same wavelength, but the reducing agent does not interfere with the tail of the O-Fe^{3+} charge transfer at 400 nm. The intensity of the absorbance at 400 nm is directly proportional to the Fe^{3+} concentration in the sample, as given by the Beer-Lambert Law,

$$A = \log I_o/I = \varepsilon lc \qquad [1]$$

where A is the absorbance; I_o and I, the incident and transmitted photon intensities, respectively; epsilon, the absorptivity; l, the sample thickness; and c, the concentration of O-Fe^{3+} moieties in the sample.

The changes in intensity at 400 nm with time of redox treatment are reported in Fig. 10–6 for soil clays. The absorbance decreased with Fe reduction due to the loss of intensity from the CT band at 245 nm as explained above. The intensity was initially high, but dropped rapidly with time after the commencement of reduction. When O_2 gas was introduced, the intensity increased steadily, reflecting the increasing level of Fe^{3+} in the sample. The final state of reoxidation recorded was below that of the unaltered sample, indicating a resistance to complete reoxidation similar to that observed in ferruginous smectite (sample SWa-1).

The CT band was also monitored continuously at 400 nm for K-saturated Fithian illite during reduction and reoxidation with the hope that some

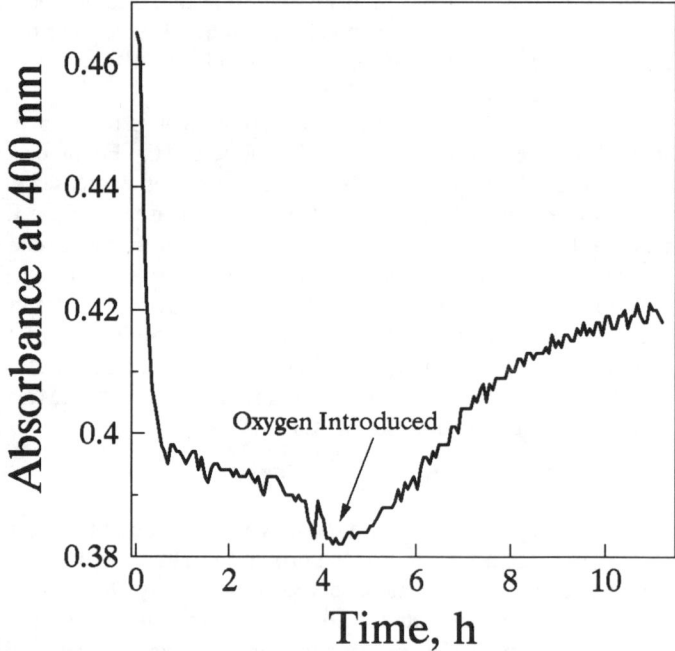

Fig. 10–6. Optical absorbance at 400 nm, deriving from O-Fe^{3+} electron charge transfer, of Drummer (Urbana) soil clay.

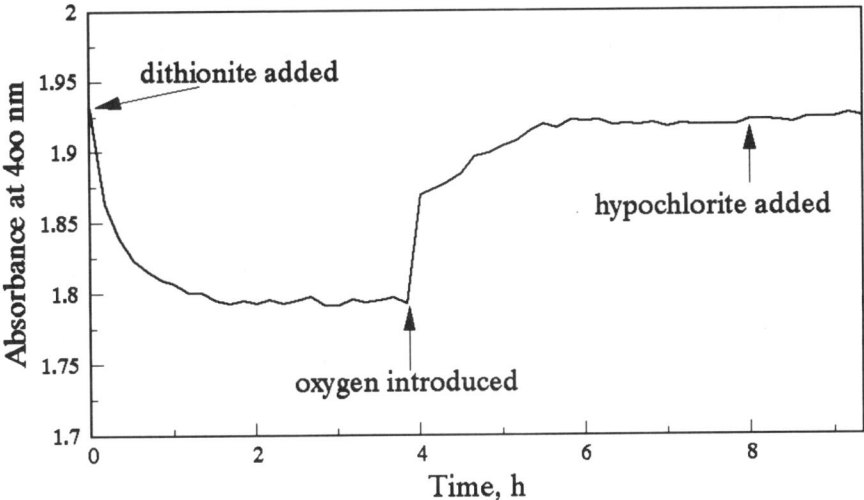

Fig. 10–7. Optical absorbance at 400 nm, deriving from O-Fe^{3+} electron charge transfer, of Fithian illite.

understanding of the behavior of illite in the soil clay fraction could be derived. Recall that in this clay 35% of the Fe initially is Fe^{2+}. When subjected to further reduction by citrate-bicarbonate-dithionite, the color changed noticeably. As illustrated by the patterns in Fig. 10–7, during the first 0.5 h of reduction of the K-saturated illite the absorbance decreased rapidly. After 3.5 h of reduction, and thereafter, the absorbance remained essentially constant. When O_2 was used to reoxidize the clay suspension, the absorbance increased quickly in ~0.5 h, then after 2 h, it returned to the level that existed before reduction. No further change in absorbance was observed despite continuous treatment with O_2, and even the introduction of sodium hypochlorite (a strong oxidizing agent) could not increase the absorbance further. These results, combined with chemical analyses (Table 10–2), signify that the original Fe^{2+} in K-saturated Fithian illite is not easily oxidized, but that which was reduced in the laboratory was rather readily reoxidized.

When the illite was reoxidized, the amount of fixed K increased and the exchangeable decreased, while the total changed only slightly. So, during one redox cycle the illite released K into its environment, in contrast to the smectite, which removed K from its surroundings.

For the Drummer soils (from Dekalb and Urbana), both fixed and exchangeable K were little changed after one redox cycle. But in the Cisne soil, by comparison, the amount of fixed K decreased and exchangeable K increased after one reduction-reoxidation step, and total K was virtually unchanged.

The trend for K behavior is, therefore, in the direction of lower K availability for smectities and greater K availability for illites after one redox cycle. The mineralogy of soils is, therefore, very important for determining the effect of Fe oxidation state on K availability in the soil. A soil containing a mixture of illite and smectite would either fix or release K during reducing conditions depending on which clay dominates; or, if these minerals are precisely balanced, no change in K availability would occur because the amount that one releases would be fixed by the other. Further study to quantify the comprehensive effects of Fe reduction on K availability for different mixtures of clay mineral types in soil clays is needed.

SUMMARY

The reduction of Fe in the crystal structures of reference and soil clay minerals alters the distribution of K between exchangeable and fixed states. In smectite clay minerals, Fe reduction increased K fixation without changing the amount available in exchangeable sites. The rate of K release was slowed and some fixed K was retained in the clay even when the sample was reoxidized. Several cyclic reduction and reoxidation steps produced increased levels of Fe^{2+} and fixed K that apparently are difficult to reverse, and these leaves may steadily increase over many such cycles. These results can be explained by the clay layers becoming collapsed as a result of Fe reduction, and thereby forming larger particles due to K fixation.

In Illite, the reduction treatment released K to the surrounding solution, which is opposite to the effect observed in smectites, possibly because of a change in electrostatic force between K and structural OH groups. Soil clay behavior was more similar to illite than to smectite. The behavior of the soil clays is thus highly dependent on the type of clay minerals present. If illite dominates, reduction may enhance K availability, whereas smectite domination will produce greater K fixation during reducing conditions.

REFERENCES

Barber, S.A. 1960. The influence of moisture and temperature on phosphorus and potassium availability. p. 435–440. *In* F.A. Van Beren et al. (ed.) Trans. Int. Congr. Soil Sci. 7th, Madison, WI. 14–24 Aug. 1960. Elsevier, Amsterdam.

Bates, T.E., and A.D. Scott. 1964. Changes in exchangeable potassium observed on drying soils after treatment with organic compounds: I. Release. Soil Sci. Soc. Am. Proc. 28:769–772.

Beckett, P.H.T. 1964. Studies on soil potassium: I. Confirmation of the ratio law: Measurement of potassium potential. J. Soil Sci. 15:1–8.

Bertsch, P.M., and G.W. Thomas. 1985. Potassium status of temperate region soils. p. 131–162. *In* R.D. Munson (ed.) Potassium in agriculture. ASA, CSSA, and SSSA, Madison, WI.

Bohannon, R.A. 1957. The effect of drying on exchangeable K in soils from Illinois and Kansas. Ph.D. thesis. Univ. of Illinois, Urbana (Diss. Abstr. AAC 0025197).

Chen, S.Z., P.F. Low, and C.B. Roth. 1987. Relation between potassium fixation and the oxidation state of octahedral iron. Soil Sci. Soc. Am. J. 41:82–86.

Doll, E.C., M.M. Mortland, K. Lawton, and B.G. Ellis. 1965. Release of potassium from soil fractions during cropping. Soil Sci. Soc. Am. Proc. 29:699–702.

Grimme, H., and K. Nemeth. 1979. The evaluation of soil K status by means of soil testing. p. 99–108. *In* Soils in Mediterranean type climates and their yield potential. Proc. Congr. Int. Potash Inst, 10th, Sevilla, Spain. 1979. Int. Potash Inst., Bern Switzerland.

Juo, A.S.R., and J.L. White. 1969. Orientation of the dipole moments of hydroxyl groups in oxidized and unoxidized biotite. Science (Washington, DC) 165:804–805.

Kaspar, T.C., J.B. Zahler, and D.R. Timmons. 1989. Soybean response to phosphorus and potassium fertilizers as affected by soil drying. Soil Sci. Soc. Am. J. 53:1448–1454.

Khaled, E.M., and J.W. Stucki. 1991. Effects of iron oxidation state on cation fixation in smectites. Soil Sci. Soc. Am. J. 55:550–554.

Komadel, P., P.R. Lear, and J.W. Stucki. 1990. Reduction and reoxidation of iron in nontronites: Rate of reaction and extent of reduction. Clays Clay Miner. 38:203–208.

Lear, P.R., and J.W. Stucki. 1985. Role of structural hydrogen in the reduction and reoxidation of iron in nontronite. Clays Clay Miner. 33:539–545.

Lear, P.R., and J.W. Stucki. 1987. Intervalence electron transfer and magnetic exchange interactions in reduced nontronite. Clays Clay Miner. 35:373–378.

Lear, P.R., and J.W. Stucki. 1989. Effects of iron oxidation state on the specific surface area of nontronite. Clays Clay Miner. 37:547–552.

McLean, E.O., J.L. Adams, and R.C. Hartwig. 1982. Improved corrective fertilizer recommendations based on a two-step alternative usage of soil tests: II. Recovery of soil-equilibrated K. Soil Sci. Soc. Am. J. 46:1198–1201.

Mengal, K., and L.C. von Branschweig. 1972. The effect of soil moisture upon the availability of potassium and its influence on the growth of many maize plants (*Zea mays L.*) Soil Sci. 134:142–148.

Mortland, M.M., K. Lawton, and G. Uehara. 1956. Alteration of biotite to vermiculite by plant growth. Soil Sci. Soc. Am. Proc. 31:286–287.

Munson, R.D. 1980. Potassium availability and uptake. p. 28–66. *In* Potassium for agriculture—A situation analysis. Potash & Phosphate Inst., Atlanta.

Munson, R.D. (ed.) 1985. Potassium in agriculture. ASA, CSSA, and SSSA, Madison, WI.

Potash and Phosphate Institute. 1980. Potassium for agriculture—A situation analysis. Potash & Phosphate Inst., Atlanta.

Reitemeier, R.F., I.C. Brown, and R.S. Holmes. 1951. Release of native and fixed nonexchangeable potassium of soils containing hydrous mica. USDA Tech. Bull. 1049. USDA, Washington, DC.

Rich, C.I. 1964. Effect of cation size and pH on potassium exchange in Nason soil. Soil Sci. 98:100–106.

Sparks, D.L., and P.M. Huang. 1985. Physical chemistry of soil potassium. p. 201–276. *In* R.D. Munson (ed.) Potassium in agriculture. ASA, CSSA, and SSSA, Madison, WI.

Sparks, D.L., and P.M. Jardine. 1981. Thermodynamics of potassium exchange in soil using a kinetics approach. Soil Sci. Soc. Am. J. 45:1094–1099.

Sparks, D.L., and W.C. Liebhardt. 1982. Temperature effects on potassium exchange and selectivity in Delaware soils. Soil Sci. 133:10–17.

Stucki, J.W., and C.B. Roth. 1977. Oxidation-reduction mechanism for structural iron in nontronite. Soil Sci. Soc. Am. J. 41:808–814.

Stucki, J.W., P.F. Low, C.B. Roth, and D.C. Golden. 1984. Effects of oxidation state of octahedral iron on clay swelling. Clays Clay Miner. 32:357–362.

Yu, T.R. 1985. Physical chemistry of paddy soils. Springer-Verlag, New York.

11 Reinventing Soil Testing for The Future

Earl O. Skogley
Montana State University
Bozeman, Montana

Soil testing's history spanning more than five decades has been plagued by many problems that defy correction. Major reasons for this are that (i) chemical extraction is not mechanistically related to plant availability of nutrients and (ii) the problems are innate to chemical extraction methodologies. A future for soil testing without these intrinsic problems will require a different approach. Ideas are presented concerning a promising alternative to chemical extraction, the background of this approach, and corroborating results from a broad array of studies. Development of this approach would allow major improvements in soil testing.

THE PROBLEM SITUATION

For purposes of this presentation, I have established 1954 as an arbitrary time of reference that relates to my first exposure to soil testing. As an undergraduate major in soils, I worked part-time in the Soil Testing Laboratory at North Dakota State College in Fargo. This was also the year of my first ASA meeting, which was the 46th annual meeting of ASA, held in St. Paul, MN, 8 to 11 Nov. 1954. There were 1492 registered members and guests; M.B. Russell was elected president of SSSA; Werner Nelson joined the Potash Institute staff; and Emil Truog was recognized as an honorary member of ISSS. There were 375 papers presented at this meeting. Several papers of personal interest addressed problems with soil testing. Major issues included accuracy, development of universal extractants, and standardization. The sodium bicarbonate extractant for P had recently been developed for use on nonacid soils (Olsen et al., 1954), and had just been adapted for soil testing in North Dakota.

The 84th annual meeting of ASA convened in Minneapolis, MN, 1 to 6 Nov. 1992. Several thousand members and guests were in attendance, and there were nearly as many papers and posters in Divisions S-4 and S-8 as the total for all soils sections in 1954. Many presentations again pertained to soil testing, with titles such as "How can the credibility of soil test-based fertilizer recommendations be improved?" and "Soil test credibility gap—who is responsible?" These titles, and hundreds of others during the intervening four decades, illustrate problems that continue to plague soil testing.

Copyright © 1994 Soil Science Society of America, 677 S. Segoe Rd., Madison, WI 53711, USA.
Soil Testing: Prospects for Improving Nutrient Recommendations, SSSA Special Publication 40.

The magnitude and duration of related problems is well illustrated by the number of soil samples that have been tested each year (Fig. 11–1). About 1.2 million samples were analyzed nationwide in 1954, and the numbers increased to ≈3.75 million in 1966 (this peak is not shown on the graph, plotted at 5-yr intervals). Since then, some overall decrease occurred and sample numbers have stabilized ≈3 to 3.5 million per year—about the same as during the early 1960s. Thus, no sustained increase in soil testing has been experienced for >30 yr.

This static level is certainly not related to *needs* for soil testing (numbers for 1993 translate into each soil sample representing an average of ≈40 ha of cropland), but rather due to the *user's perception of the value* of soil testing. It is generally accepted that credibility *is* a serious problem (just recall the two paper titles from 1992 that were quoted above). The problems that have been so thoroughly discussed during the past few decades, and the inability to solve these problems, are responsible for this lack of soil test credibility.

Are these problems real? Discussing the *same* problems, decade after decade, proves that they *are* real. Today's program proves that these problems have still not been solved and *continue* to plague soil testing. Furthermore, for chemical-extraction based soil testing, the same problems will be talked about for *decades to come*. These problems are *intrinsic* to this approach to soil testing.

Is there a solution? I believe that the innate nature of these problems translates into an absolute need for a change in approach if satisfactory solutions are to be realized. The nation has been asked to reinvent quite a few things during the past couple decades, including the car, government, and our medical care system. As with these other flaws in our society, solving soil test problems will require reinventing soil testing.

The theme of this, the 85th Annual Meeting of our Societies, is *Building Bridges*. It is in this spirit "looking out to the future, sharing visions of what might be" that I present these ideas for a soil testing future with fewer, solvable problems.

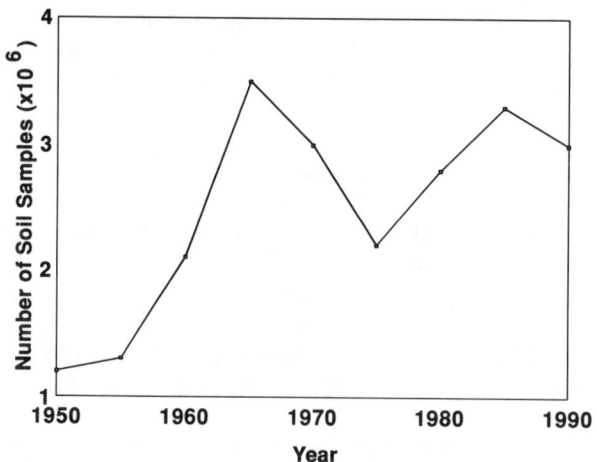

Fig. 11–1. Number of soil samples: USA (from Soil-Plant Analyst 9(2), 1991).

THE IDEAL SOIL TEST

The underlying premise of soil testing is that a representative soil sample can be obtained, tested in some way, and values derived for different nutrients that provide the basis for a fertilizer recommendation for the field (and crop) represented by the soil sample. To do this effectively, several general characteristics of the process must be considered. The following list of characteristics for a laboratory procedure would provide something pretty close to an *ideal* soil test:

1. Extract all nutrients simultaneously (*universal*).
2. Work on all kinds of soils (*standardizable*).
3. Be *accurate*.
4. Be *simple*.
5. Be *cost-effective*.
6. Be *rapid*.

History tells us that soil testing based on chemical extraction methodologies can assure only Point 6 (rapidity) on this list. Some degree of cost-effectiveness may also be claimed. Most of the other characteristics of an ideal soil test have been sacrificed for the sake of rapidity, and even this is in question under certain conditions. After so many years, tradition now also plays a large role in the way soil testing is done.

The quest for a single chemical extractant that will be useful for the entire array of plant nutrients (universal extraction) will probably never end. Furthermore, chemical extraction methodologies will not allow standardization. Even within a given area, soils with certain properties mandate the use of different extractants.

Of course, results obtained from the laboratory analysis of the soil sample must somehow relate to how the specific crop will respond to specific fertilizers applied to the field from which the sample was collected. This requires correlation and calibration studies, which is another area of weakness for today's soil testing situation, and accuracy is compromised due partly to this weakness.

Current methodologies certainly are not simple; numerous steps are required, each one providing opportunities for error. This, in turn, reflects on the accuracy of results.

Additionally, accuracy is limited because the approach is not based directly on factors that control nutrient availability. Because of this, I believe it is necessary to add a seventh characteristic to the list for an ideal soil test:

7. That it be based on sensitivity to mechanisms that control nutrient availability to plants.

Without this as a base, it is doubtful that many of the other ideal characteristics could be realized.

Several attempts have been made to develop a methodology that provides desired soil test characteristics. For example, van Raij (1994) from Brazil states that the ideal *universal extractant* should be (i) rapid, (ii) reproducible, (iii) economical, (iv) adapted to soils from different regions, and (v) a measure of labile forms that supply plant roots. He also states that "Most extractants in use fall short of these requirements and are in reality *multinutrient* extractants that often give priority to the laboratory convenience." van Raij has developed a soil test methodology based on ion-exchange resins that fulfills most of these criteria, but it is not very simple, requires more time for extraction than current methods, and is not sensitive to nutrient diffusion, a major controlling factor for nutrient availability. In spite of these limitations, sample numbers in the state of Sao Paulo, Brazil, went from 65 000 in 1982 (the test was adopted in 1983) to 250 000 in 1989 (van Raij et al., 1994). This 385% increase is probably related to the public's perception of value received.

How can soil testing be *reinvented*? Results from studies during the past several years suggest that ion-exchange resins provide a solid foundation on which to build. Ion-exchange resins were developed during the 1940s. In spite of the fact that they share many characteristics with soils, they have received little attention by most soil scientists. Tremendous advances have been made in the manufacture of resins, and they are now available with a broad array of specific characteristics (Dorfner, 1991). By taking advantage of selected properties of resins, many advances in soil science can be attained, including an improved soil test methodology.

A PROPOSED METHODOLOGY

Scientists at Montana State University have studied a methodology (based on ion-exchange resins) that provides the basis for a soil test that can fulfill each of the seven characteristics of an ideal soil test. It is important to point out from the start, however, that if Point 7 is met (that the methodology be based on mechanisms that control nutrient availability), then it is likely that a successful soil test may not be as *rapid* as some would prefer. Ion diffusion is a primary mechanism that controls nutrient availability at the plant root surface, and an *instantaneous* measure of this process in soils may be difficult to obtain.

7. *Mechanistic Relation to Nutrient Availability*: The basis of this technology is the encapsulation of mixed-bed ion-exchange resins. Each capsule is spherical, with a rigid surface that is porous and of uniform area. These properties provide a system in which the area of contact between the soil and the adsorber is known and remains constant during the *extraction* period. The quantity of resin in the capsule is large enough to act as an effectively infinite *sink* for ions for long periods in most media. The quantities of solutes accumulated in the capsule during a specified time are dependent upon *solute concentration* and *rate of solute diffusion* to the sink (Yang et al., 1991a,b; Yang & Skogley, 1992).

An example of the relationship between resin adsorption of nutrients and soil solution concentration of those nutrients is presented in Table 11–1. This relationship was highly significant for K, P, S, and NH_4. Each of these nutrients is known to be highly dependent upon diffusion for its movement to plant roots.

Data in Table 11–2 provide an example of the influence of nutrient ion diffusion through the soil as a controlling factor to resin adsorption of these nutrients. The excellent relationships between fractional attainment of adsorption and time for K, P, and S by resin capsules indicates that diffusion of these ions through the soil is rate limiting in resin adsorption.

These results indicate that soil properties in direct control of nutrient availability to plant roots are the same properties that determine nutrient quantities

Table 11–1. Regression equations for the relationship between rate constant (k)† for nutrient accumulation (µmol cm^{-2}) in resin capsules and respective nutrient ion activity (mM) in soil solution‡ in three Montana soils (Yang & Skogley, 1992).

Equations			r^2
k (RAQ K)§	=	$0.012 + 0.485 \propto K$	0.82*
k (RAQ P)	=	$-0.011 + 1.180 \propto HPO_4$	0.85*
k (RAQ S)	=	$-0.019 + 0.174 \propto SO_4$	0.96*
k (RAQ NH–N)	=	$-0.006 + 0.179 \propto NH_4$	0.86*

* Significant at the 0.01 level of probablity.
† From Parabolic Diffusion Model.
‡ Extracted by immiscible liquid displacement.
§ RAQ, resin adsorption quantity (µM cm^{-2} of capsule surface).

Table 11–2. Regression equation for fractional attainment (A) of adsorption of K, P, and S by resin capsule during 1 to 7 d (t) as compared with 15 d of equilibration in five soils (Yang et al., *1991*).

Soil	Equation†	R^2
Tanna	$A(t)_K$‡ = 0.16 + 0.08t	0.96*
	$A(t)_P$ = 1.28 + 0.15t	0.99**
	$A(t)_S$ = 0.04 + 0.19t	0.93*
Fort Collins	$A(t)_K$ = 0.04 + 0.09t	0.99**
	$A(t)_P$ = 0.01 + 0.12t	0.96*
	$A(t)_S$ = 0.02 + 0.10t	0.89*
Rothiemay	$A(t)_K$ = 0.08 + 0.10t	0.99**
	$A(t)_P$ = 0.12 + 0.10t	0.99**
	$A(t)_S$ = 0.12 + 0.11t	0.98**
Kevin	$A(t)_K$ = 0.11 + 0.08t	0.99**
	$A(t)_P$ = 0.21 + 0.05t	0.99**
	$A(t)_S$ = 0.13 + 0.10t	0.93*
Stryker	$A(t)_K$ = 0.50 + 015t	0.90*
	$A(t)_P$ = 0.13 + 0.06t	0.96*
	$A(t)_S$ = 0.14 + 0.25t	0.90*

*,** Significant at the 0.05 level and 0.01 probability levels, respectively.
†$A(t)_i$ = -log [1-(Q_t/Q_{15})], where t = 1, 3, 5, and 7 d, and the unit for Q is µmol cm^{-2}, assuming 15-d extraction is near quasiequilibrium.

accumulated by these resin capsules. Thus, Point 7 (that the method be sensitive to mechanisms that control nutrient availability to plants) for an ideal soil test is achieved. Building on this *mechanistic* relationship, the approach has been evaluated in relation to the other desired characteristics of an ideal soil test.

1. Universal Extraction: Using a mixed-bed resin system ensures that cations and anions are both adsorbed. The resins are strong-base and strong-acid, and initially saturated with H$^+$ and OH$^-$. This ensures that the resins have a greater affinity for all other ions (except bicarbonate) in the surrounding medium and will continue to function as a sink (or extractant) during the testing period. Nutrient uptake by these capsules is simultaneous and independent (Yang, 1991a; Fig. 11–2). In this manner, Point 1 (universal extraction of nutrients) can be accomplished.

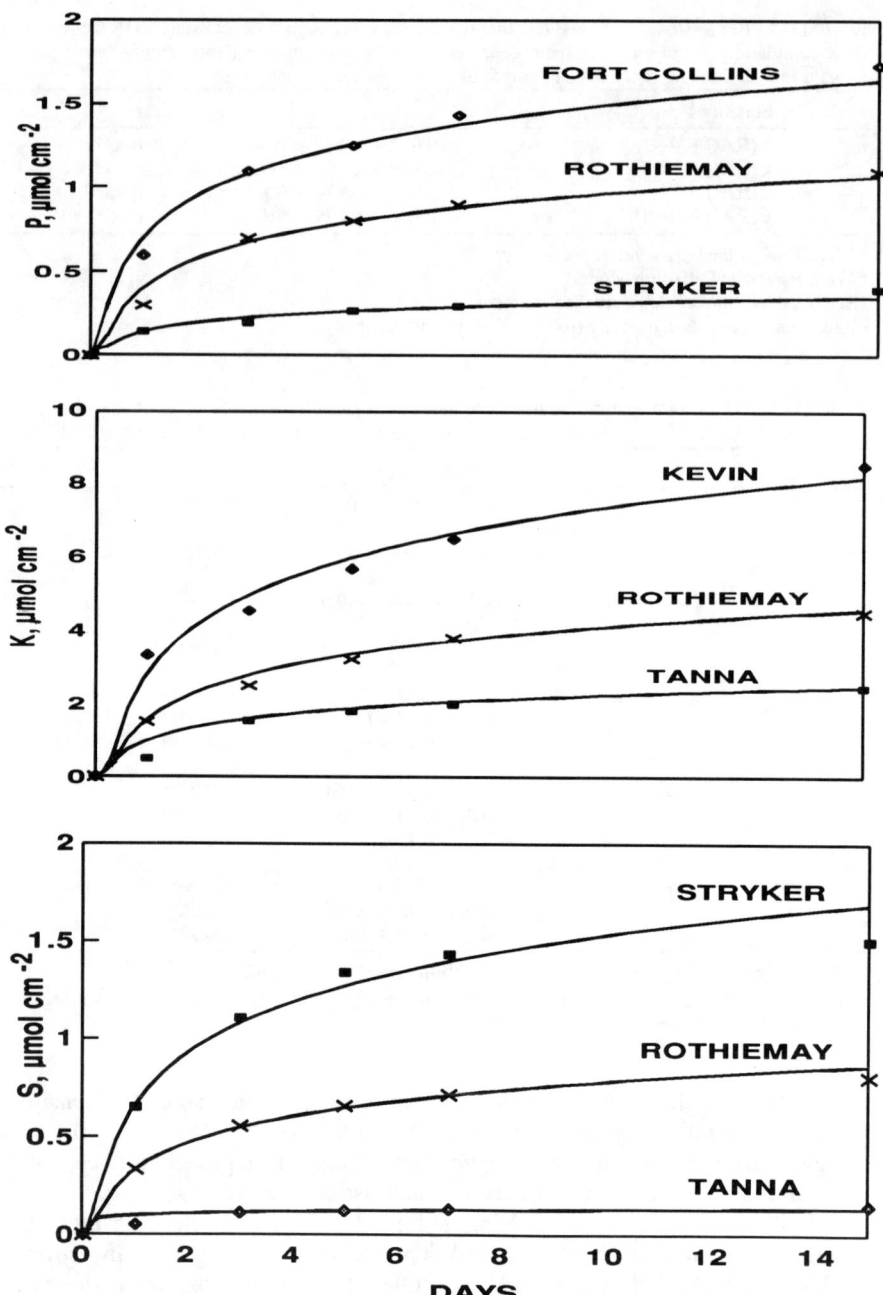

Fig. 11–2. Nutrient uptake curves with resin capsules for representative high, medium, and low nutrient-supplying soils (from Yang et al., 1991a).

2. *Standardization*: Soil samples representing 10 soil orders from around the world have been tested with resin capsules. Nutrient release curves have been obtained for a broad array of elements on each of these soils, indicating that the technology will function on soils of virtually all kinds (Fig. 11–3). There is no need to change the extractant when soils of different natures are tested. Thus, the basis for standardization, worldwide, is provided.

3. *Accuracy*: Improved soil test accuracy is made possible with this approach in several ways. Results are based on sensitivity to mechanisms that control nutrient availability. Sample handling steps (e.g., drying and grinding) that are known to alter nutrient extractability are eliminated. The procedure is simple, requiring very few steps, so opportunities for error are minimized. Uniform protocols could be developed that would provide the best opportunity for accurate soil testing.

4. *Simplicity*: Few steps are required with this methodology. The procedure involves adding water to the field-fresh soil sample to make it into a saturated

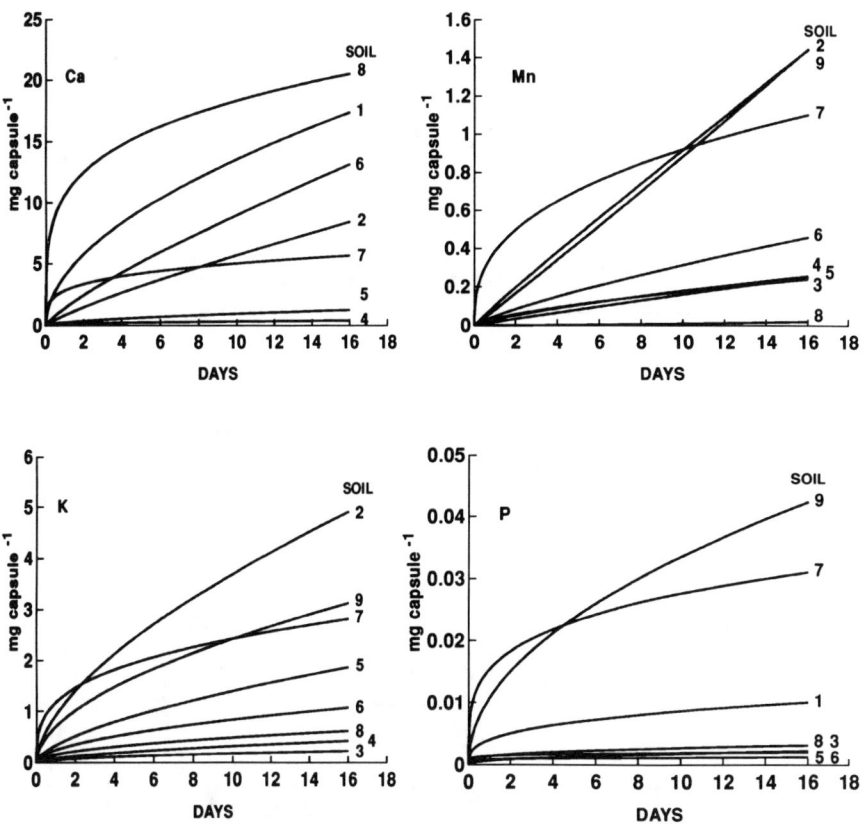

Fig. 11–3. Typical resin capsule nutrient uptake curves from soils representing seven soil orders from throughout the world (A. Dobermann, 1993, personal communication). Soil: 1, Russia; 2, Cuba; 3, Vietnam; 4, Philippines; 5, Nicaragua; 6, Ethiopia; 7, Germany; 8, Syria; 9, Germany.

paste, and mixing it thoroughly. This provides sample uniformity and a standard soil-water status, as well as conditions for maximum rate of diffusion. Little or no error is created if exact saturated paste conditions are not attained (Fig. 11–4; Skogley et al., 1990). A subsample of the paste is placed into a container and a resin capsule inserted so that it is completely surrounded by a few millimeters of soil paste (Fig 11–5). The container is capped and set aside at a standard temperature for a specified time. As mentioned earlier, the time allowed must be sufficient to allow effects of ion diffusion to be expressed. At the end of this adsorption period, the capsule is removed from the paste, rinsed with pure water to remove any adhering soil particles, and desorbed with 50 mL of 2 M HCl (or other desired reagent) at the rate of ≈1 mL min^{-1}. The resultant analyte contains

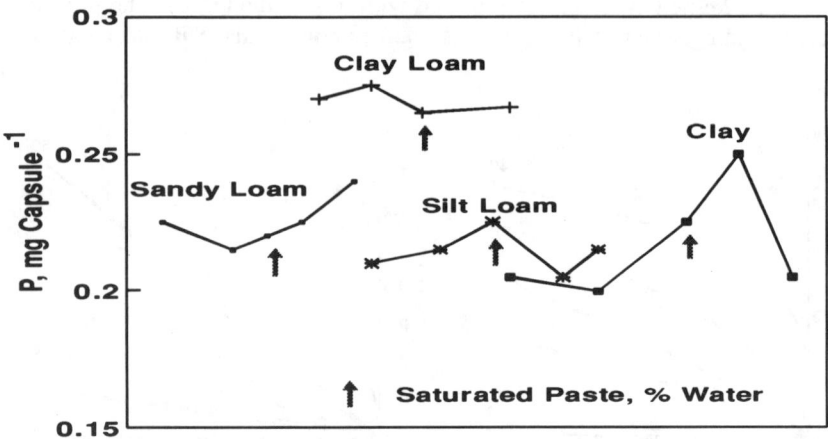

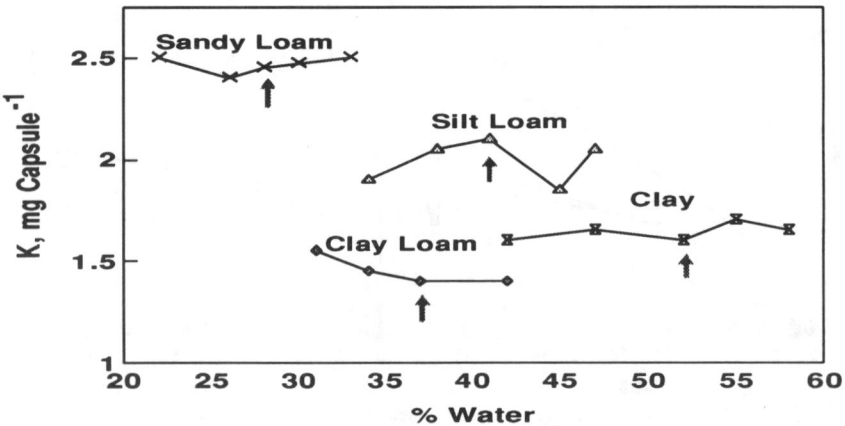

Fig. 11–4. Effect of deviations above or below saturated paste soil water conditions on nutrient accumulation by resin capsules.

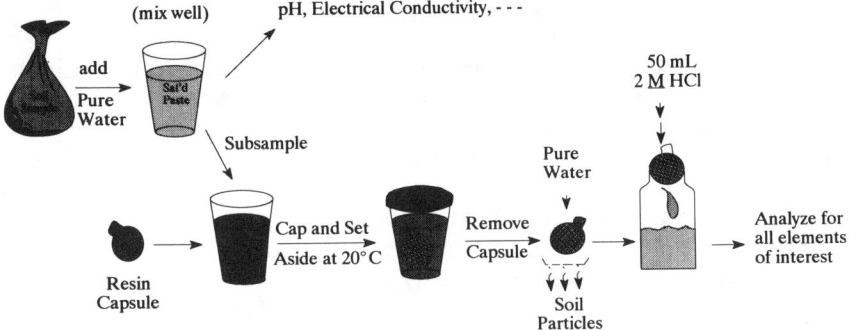

Fig. 11–5. Schematic flow diagram for resin capsule soil test procedure.

most of each element that was accumulated by the resin capsule and can be analyzed by standard procedures.

That portion of the sample not used for resin extraction can be used to determine soil pH, electrical conductivity, texture, organic matter, or other desired analyses. Many of these procedures for saturated paste analysis are already in use. Also, perhaps relationships can be developed from results of resin capsule analysis that would provide good estimates of some of these sample properties.

5. Cost-Effectiveness: The cost-effectiveness of this approach to soil testing remains to be determined. Elimination of many manpower, chemical, space, labware, and energy intensive steps, however, allows for reduced laboratory costs. Results that are more accurate than those from current soil tests certainly would provide a large potential cost-effectiveness advantage.

Additional savings could be realized by making saturated pastes and extracting samples at the site of collection (e.g., by those taking the soil sample) and sending only the resin capsule to the laboratory. About 100 capsules could be shipped for the cost of a single 0.5-kg soil sample.

6. Rapidity: An extraction period extending during several days may be too long for routine soil test application. A standard period would have to be established, based on a broad cross-section of results. This may turn out to be only a day or two, and a portion of this extraction period is compensated by using field-fresh samples. Drying, grinding, and numerous other time-intensive steps required by chemical extraction are eliminated. Thus, time requirements may be no more than in cases where samples are extracted more than once to obtain a complete nutrient analysis by current methods.

Perhaps short-cuts can be developed to reduce the extraction period. One such approach would be to initially develop nutrient release curves, spanning several days, for specific soils or groups of soils. Subsequent tests on these soils could then be done at some abbreviated time to provide the current nutrient status of that soil. Also, for many soils, it may be possible to infer differences in nutrient diffusion from other immediately obtainable or already known soil properties. Nevertheless, it is acknowledged that the trade-off for an improved soil test *may* be an extra day or two in the laboratory.

PROVING THE METHODOLOGY

Numerous studies have been conducted to investigate various aspects of this resin capsule methodology. Research on K requirements for Montana soils and crops (Skogley, 1975), as well as a history of experience with ion-exchange resins, led to this approach. In a statewide soil fertility program initiated in 1972, crops frequently responded to fertilizer K on soils that tested high in ammonium acetate extractable K (Fig. 11-6). Attempts to develop an improved laboratory soil test for K revealed that a resin extraction approach provided the best relationship to these crop responses (Haby, 1975).

Meanwhile, it had been demonstrated that the most important soil properties and processes controlling the availability of K at the plant root surface are (i) those that determine K concentration in the soil solution, which directly influences amounts of K swept to roots by mass-flow and (ii) the rate of diffusion of K toward the root (Barber, 1962; Beckett, 1964; Thomas & Hipp, 1968). Contributions of mass-flow and diffusion to availability of nutrients other than K have also been reported over the years (Barber, 1984). Thus, it became obvious that an improved test for many nutrients on a wide range of soils would probably need to measure, or at least closely relate to, these parameters that control nutrient availability.

Many years of experience with using ion-exchange resins to study various soil and plant nutrient relationships provided the knowledge and background to further investigate their potential use for soil testing (Skogley, 1966, 1969; Skogley & Dawson, 1963; Skogley & Haider, 1969; Leite & Skogley, 1977; Schaff & Skogley, 1982, Skogley & Schaff, 1985). Other researchers reported results of diverse resin-soil studies, providing a sound basis on which to proceed (see Dobermann et al., 1994, for an extensive literature review on this topic).

As described above, it was predicted that a resin capsule system could provide the desired characteristics for an improved K soil test. At the same time, it

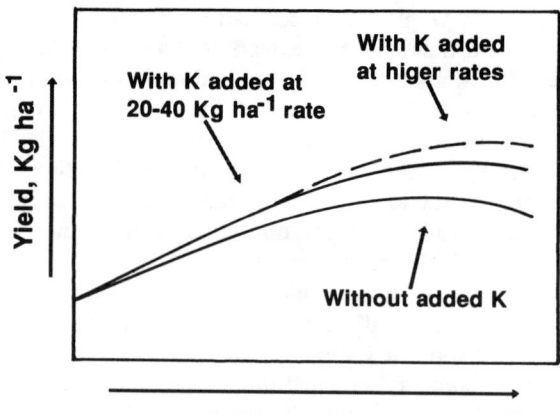

Fig. 11-6. Typical small grain response to N and P fertilizers with or without K fertilizer on K responsive sites in Montana.

was obvious that other cations would also be adsorbed by the resin. Further, by using a capsule filled with mixed-bed resin, the test could be expanded to include anions as well as cations. Studies by Georgitis (1989) proved that this approach provided the basis for a soil test that related well to nutrient uptake by plants, and for a cation (K) as well as for anions (S and P). Subsequent studies provided basic descriptions of system dynamics, kinetics, and general function (Skogley et al., 1990; Yang et al., 1990a,b; 1991a,b; Yang & Skogley, 1992; Skogley, 1992).

The characteristics of the resin capsule system also provided an approach for in situ testing and monitoring. Studies have been conducted to determine resin capsule utility for detecting and quantifying solute transfer in the field and laboratory, and as influenced by various conditions. Results have proven that this approach can be highly sensitive, and could be developed to provide more accurate, cost-effective results for field studies compared with current methodologies (Montagne & Skogley, 1992; Li et al., 1993). The approach is also being expanded to include testing and monitoring of organic substances (Johns et al., 1993; Johns & Skogley, 1994).

Several side-by-side comparisons between this methodology and current soil tests are being conducted. A study is under way in which several methods (including resin capsules) for P extraction are being compared by 10 laboratories throughout the USA. This study is under the direction of Dr. Andrew Sharpley, USDA-ARS, Durant, OK. Results should be available in the next several months. Dr. Paul Fixen, Phosphate and Potash Institute, directed a study in which >80 soil samples were obtained from corn (*Zea mays* L.) fields in Minnesota, Iowa, and South Dakota during the 1992 growing season. Soil samples and corn tissue samples at near the four-leaf stage were collected from the same locations and analyzed. Soil samples were split, with one portion being analyzed by current methods in two Midwest soil testing laboratories. Another portion was tested with resin capsules. Correlations between standard and resin capsule test results for P and K are presented in Table 11–3. These data clearly illustrate that resin capsule results relate to other soil test values just as well as do the various current tests among themselves.

All correlations between soil test values and plant P and K concentrations were poor, but again the resin capsule method appears to have been at least as sensitive to factors that influence P and K uptake by corn as were the standard methods (Table 11–4).

Table 11–3. Correlations among selected soil test† values on samples collected from corn fields in Iowa, Minnesota, and South Dakota, May 1992 ($n = 81$).

P Soil Test	R^2	K Soil Test	R^2
BP vs. OP	0.76**	K_{Ext} (Lab 1 vs. Lab 2)	0.92**
BP vs. RP	0.58**	K_{Ext} vs. K_{Rel}	0.56**
OP vs. RP	0.80**	K_{Ext} vs. K_{Res}	0.82**
		K_{Rel} vs. K_{Res}	0.62**

** Significant at the 0.01 probability level.
†BP = Bray P; OP = Olsen P; RP = Resin P (2 d); K_{Ext} = NH$_4$ OAC K; K_{Rel} = K release; K_{Res} = Resin K (2 d).

Table 11–4. Correlations among selected soil test[†] values and P or K concentrations (dry wt. basis) of corn at about the four-leaf stage. Samples from Iowa, Minnesota, and South Dakota ($n = 81$).

P soil test	R^2	K	R^2
BP vs. corn P	0.006	K_{Ext} vs. corn K	0.243
OP vs. corn P	0.019	K_{Rel} vs. corn K	0.266
RP vs. corn P	0.030	K_{Res} vs. corn K	0.277

[†] BP = Bray P; OP = Olsen P; RP = Resin P (2 d); K_{Ext} = NH_4 OAC K; K_{Rel} = K release; K_{Res} Resin K (2 d).

Another comparison of methods is being conducted in a commercial soil testing laboratory in Montana. Soil samples from a broad area of the Northern Great Plains and Intermountain Valleys of the USA and Canada are being analyzed by standard and resin capsule methods. Results will provide an indication of relative accuracy for predicting fertilizer responses for a wide variety of soils and cropping conditions in actual farming situations. Comparative laboratory costs will also be determined.

A rigorous test of the resin capsule technology was conducted by German scientists on a worldwide collection of soil samples representing seven soil orders. Resin capsule data were collected at 1 to 16 d of adsorption for P, K, Na, Ca, Mg, Fe, and Mn. Representative results are shown in Fig. 11–3. Typical nutrient release curves were obtained for each of these soils, with large differences being expressed. Simultaneous, independent adsorption of nutrients was demonstrated on each soil, representing extreme diversity in properties from widely separated world regions.

This research was expanded to soils used for flooded rice production in Pacific Rim countries (Dobermann et al., 1993). Nutrient release curves were developed for 14 d periods for N, P, K, Na, Ca, and Mg. Again typical curves were obtained, with large differences being expressed for different soils (Fig. 11–7).

An example of resin capsule sensitivity to past fertilizer history is presented in Fig. 11–8. At this location where long-term fertilizer treatments had been applied annually to rice (*Oryza sativa* L.), the resin adsorption kinetics for P clearly expressed effects of the fertilizer treatments. Soils that had received P annually had much greater resin-adsorbed P than those that received no P. There was a striking build-up of P reserves in plots that had received only P, reflecting lower uptake rates due to N or K deficiencies, as compared with the N–P–O or N–P–K treatments. The uptake of P by rice would have been greatest from the N–P–K plots, so the accumulation of P would be only moderate under this treatment. This result was clearly expressed by the resin data. Where no P was applied, results clearly indicated this, even to the point of showing reduced P availability where N or K had been applied and would have resulted in somewhat increased plant uptake of native soil P.

It is clear from results of basic studies that the resin capsule methodology functions in response to soil characteristics and processes that regulate nutrient availability. Additional support for this approach is provided by positive results from its use on soils of nearly all natures, and under many field and laboratory conditions. Thus, a solid basis has been established for a soil test that does not harbor the built-in problems of the chemical extraction approach.

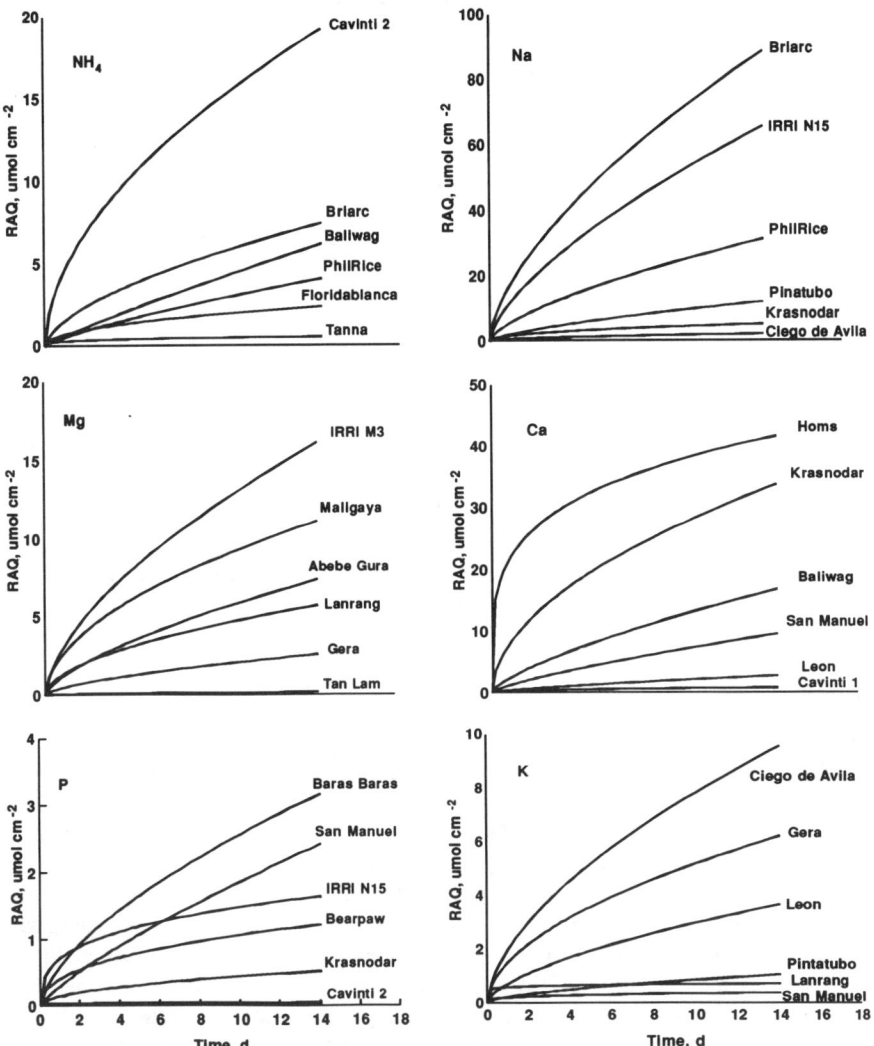

Fig. 11–7. Resin adsorption of nutrients in various Asian rice soils, with two Montana soils (Bearpaw and Tana) for comparison (Dobermann et al., 1994).

FUTURE DEVELOPMENT

A soil test future without any problems is certainly a dream. An improved soil test that provides universal extraction and that could be standardized for all areas of the world, however, is a real possibility. This will not be quick or easy, and it certainly cannot be accomplished within any single local program. A nationwide, cooperative effort, with involvement from many states or regions is necessary. Due to drastic divergence from chemical extraction, correlation of resin capsule results with those from current methods will provide only a general

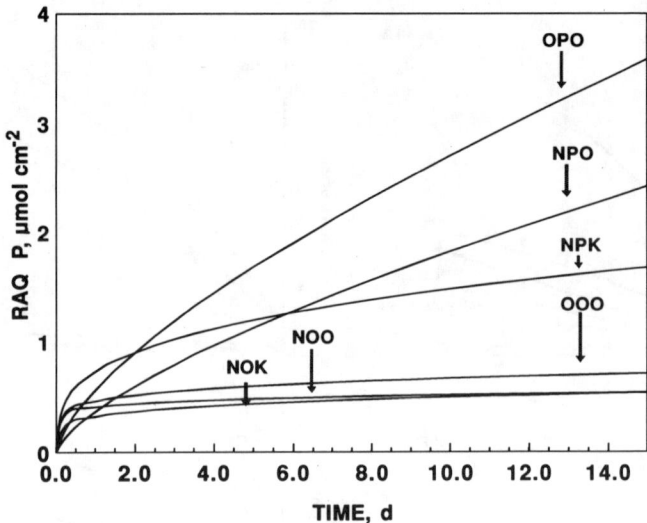

Fig. 11–8. Phosphorus adsorption by resin capsules in six treatments of a long-term fertility trial with rice (Doberman et al., 1994). NPK refers to annual application of nitrogen, phosphorus, and potassium. O indicates that no fertilizer of that type was added.

data base for comparison. Improving accuracy to the extent made possible by this new approach will require correlation and calibration studies under field conditions.

Although correlation and calibration field studies are not as numerous and extensive now as they were during the formative years of soil testing, several state and regional sample testing and fertilizer recommendation programs are currently active. These activities could quickly provide a solid data base for a new methodology if all results were combined in a unified, national program. Uniform, simplified testing protocols could be developed simultaneously. Nationwide standardization could soon become a reality. A comprehensive computer program could be developed to include appropriate algorithms for fertilizer recommendations for different crops, as influenced by appropriate inputs, and including environmental considerations. Expansion to include foreign countries and international agencies could develop the foundation for standardization on a world-wide scale.

Development of this new methodology could allow soil testing to proceed into the next century without the burden of problems innate to methodologies based on chemical extraction of soil samples. The footings for the bridge are in place, can the bridge to the future be built?

REFERENCES

Barber, S.A. 1962. A diffusion and mass-flow concept of soil nutrient availability. Soil Sci. 93:39–49.
Barber, S.A. 1984. Soil nutrient bioavailability—A mechanistic approach. Wiley Interscience, New York.
Beckett, P.H.T. 1964. Studies on soil potassium: II. The immediate Q/I relations of labile potassium in the soil. J. Soil Sci.15:9–23.

Dobermann, A., H. Langner, H. Mutscher, J.E. Yang, E.O. Skogley, M.A. Adviento, and M.F. Pampolino. 1994. Nutrient adsorption kinetics of ion exchange resin capsules: A study with soils of international origin. Commun. Soil Sci. Plant Anal. 25:1329–1353.

Dorfner, K. (ed.). 1991. Ion exchangers. Walter de Gruyter Berlin, New York.

Georgitis, S.J. 1989. The development and characterization of the phytoavailability soil test for potassium, sulfur, and phosphorus. Ph.D. diss. Montana State Univ., Bozeman.

Haby, V.A. 1975. Evaluation of soil test methods and development of a potassium fertilizer recommendation system for Montana. Ph.D. diss. Montana State Univ., Bozeman (Diss. Abstr. 75-25,754).

Johns, M.M., and E.O. Skogley. 1994. Soil organic matter testing and labile carbon identification by carbonaceous resin capsules. Soil Sci. Soc. Am. J. 58:751–758.

Johns, M.M., E.O. Skogley, and W.P. Inskeep. 1993. Characterization of carbonaceous adsorbents by soil fulvic and humic acid adsorption. Soil Sci. Soc. Am. J. 57:1485–1490.

Leite, J.P., and E.O. Skogley. 1977. Retention and leaching of copper and zinc in "Tabuleiro" soils as influenced by nutrient carrier. Pesqui. Agropecu. Bras. 12:27–34.

Li, Z.M., E.O. Skogley, and A.H. Ferguson. 1993. Resin adsorption for describing bromide transport in soil under continuous or intermittent unsaturated water flow. J. Environ. Qual. 22:715–722.

Montagne, C., and E.O. Skogley. 1992. Burned-unburned resin extractor soil nutrient comparison: Yellowstone National Park. Final Rep. (mimeo) to USDA Natl. Park Serv., Mammoth, WY. Montana State Univ., Bozeman.

Olsen, S.R., F.S. Watanabe, and L.A. Dean. 1954. Estimation of available phosphorus in soils by extraction with sodium bicarbonate. Circ. 939. U.S. Gov. Print. Office, Washington, DC.

Schaff, B.E., and E.O. Skogley. 1982. Diffusion of potassium, calcium and magnesium in Bozeman silt loam as influenced by temperature and moisture. Soil Sci. Soc. Am. J. 46:521–524.

Skogley, E.O. 1966. Ion-exchange resin media: Micronutrient levels and the response of tomatoes. Soil Sci. 102:167–172.

Skogley, E.O. 1969. The "Donnan Theory" in development of plant growth media from ion-exchange resins. Agron. J. 61:317–322.

Skogley, E.O. 1975. Potassium in Montana soils and crop requirements. Res. Rep. 88. Montana Agric. Exp. Stn., Montana State Univ., Bozeman.

Skogley, E.O. 1992. The universal bioavailability environment/soil test: UNIBEST. Commun. Soil Sci. Plant Anal. 23:2225–2246.

Skogley, E.O., and J.E. Dawson. 1963. Synthetic ion exchange resins as a medium for plant growth. Nature (London) 198:1328–1329.

Skogley, E.O., S.J. Georgitis, J.E. Yang, and B.E. Schaff. 1990. The phytoavailability soil test-PST. Commun. Soil Sci. Plant Anal. 21:1229–1243.

Skogley, E.O., and S.S. Haider. 1969. Soil-resin system studies: Effects of sodium and magnesium on barley (*Hordeum vulgare* L.) and tomatoes (*Lycopersicon esculentum* Mill.). Plant Soil 30:343–359.

Skogley, E.O., and B.E. Schaff. 1985. Ion diffusion in soils as related to physical and chemical properties. Soil Sci. Soc. Am. J. 49:847–850.

Thomas, G.W., and B.W. Hipp. 1968. Soil factors affecting potassium availability. p. 269–291. *In* V.J. Kilmer et al. (ed.) The role of potassium in agriculture. ASA, CSSA, and SSSA, Madison, WI.

van Raij, B. 1994. New diagnostic techniques, Universal Soil Extractants. Commun. Soil Sci. Plant Anal. 25:799–816.

van Raij, B., H. Cantarella, J.A. Quaggio, L.I. Prochnow, G.C. Vitti, and H.S. Pereira. 1994. Soil testing and plant analysis in Brazil. Commun. Soil Sci. Plant Anal. 25: 739–751.

Yang, J.E., E.O. Skogley, and B.E. Schaff. 1990a. Microwave radiation and incubation effects on resin-extractable nutrients: I. Nitrate, ammonium, and sulfur. Soil Sci. Soc. Am. J. 54:1639–1645.

Yang, J.E., E.O. Skogley, and B.E. Schaff. 1990b. Microwave radiation and incubation effects on resin-extractable nutrients: II. Potassium, calcium, magnesium, and phosphorus. Soil Sci. Soc. Am. J. 54:1646–1650.

Yang, J.E., E.O. Skogley, S.J. Georgitis, B.E. Schaff, and A.H. Ferguson. 1991a. The phytoavailability soil test: Development and verification of theory. Soil Sci. Soc. Am. J. 55:1358–1365.

Yang, J.E., E.O. Skogley, and B.E. Schaff. 1991b. Nutrient flux to mixed-bed ion-exchange resin: Temperature effects. Soil Sci. Soc. Am. J. 55:762–767.

Yang, J.E., and E.O. Skogley. 1992. Diffusion kinetics of multinutrient accumulation by mixed-bed ion-exchange resins. Soil Sci. Soc. Am. J. 56:408–414.

12 Current Approaches to Soil Testing Methods: Problems and Solutions

E.R. Allen and G.V. Johnson
Oklahoma State University
Stillwater, Oklahoma

L.G. Unruh
Texas A&M University
College Station, Texas

Methods of soil testing have been evolving since testing first began. The traditional approach has been to extract the soil with a solution that quickly and reproducibly removes a nutrient fraction that is easily measured analytically and is correlated with crop response. Even though this approach has proven successful and remains the most feasible approach to date, problems persist. A major problem is diversity in the type of extractants used across the country. Debate over the best extractants for a particular area spurs much controversy. Many laboratories resist changing soil test methods because of loyalty to researchers or institutions that developed particular methods, or because of difficulties (time or expense) associated with correlating and calibrating new extractants and implementing new procedures in established laboratories. Soil testing controversies, and associated bickering surrounding appropriate soil test methods, cause confusion and a loss of credibility in the eyes of farmers and the public. As a result, the use of soil testing as a management tool is undermined even though it is one of the most cost effective practices available to farmers. This chapter reviews current approaches to soil testing for N, P, and K. Selected extractants are proposed for adoption nationally as standardized soil test methods. National standardization of soil test methods will improve the credibility of soil testing and lead to its more extensive use.

HISTORICAL DEVELOPMENT OF SOIL TEST METHODS

A better appreciation of the current approach to soil testing can be gained by considering historical development. Selected highlights are presented in Table 12–1. Columella described what may be considered the first soil test in 50 B.C.

Copyright © 1994 Soil Science Society of America, 677 S. Segoe Rd., Madison, WI 53711, USA.
Soil Testing: Prospects for Improving Nutrient Recommendations, SSSA Special Publication 40.

Table 12–1. Selected historical highlights in soil testing.

Date	Location	Event
50 B.C.	Rome	Columella recommends taste test to measure acidity and salinity
1842 A.D.	Germany	Liebig states his law of the minimum
1845	England	Daubeny describes *active* and *dormant* soil nutrient fractions and extracts them with carbonated water
Late 1800s	USA	Hilgard promotes strong hydrochloric acid for determining soil fertility status
1894	England	Dyer uses 1% (v/v) citric acid as a P and K soil test
1909	Germany	Mitscherlich develops relationship between plant growth and nutrient supply
Early 1900s	USA	Hopkins promotes monitoring of soil fertility status to avoid nutrient depletion
1930	USA	Chapman and Kelly use neutral ammonium acetate as extractant for exchangable bases
1940s and 1950s	USA	Crop varieties and fertilizer availability spur interest in soil testing as a management tool
1940s–1980s	USA	Smorgasbord of multi-element and single-element extractants developed
1970s to present	USA	Standardization of methods pursued and alternative approaches investigated

He estimated the acidity and salinity status of soil by tasting it (Tisdale et al., 1985). At the start of contemporary soil science, agricultural chemists believed soil fertility status could be evaluated by measuring total soil nutrient content (Viets, 1980). This approach was soon found to be inadequate since total nutrient content often did not correlate well with nutrient availability. A more modern approach to soil testing began in the mid 1800s when general principles of plant nutrition, such as Justis von Liebig's law of the minimum, were gaining some acceptance. A good review of the development of soil testing from the 1840s through the 1950s is provided by Anderson (1960). In his review, he divided the progression of soil testing into three loosely defined periods.

The first period described by Anderson (1960) was from 1845 to 1906 and dealt primarily with the work of four scientists who laid the foundation for the modern approach to soil testing, C.G.B. Daubeny, J. von Liebig, E.W. Hilgard, and B.J. Dyer. All four developed or evaluated rapid extraction methods for estimating soil fertility status. Daubeny (1845) suggested the terms *active* to describe more soluble nutrients and *dormant* to describe less soluble nutrients. He used carbonated water to extract active fractions, but analytical difficulties of that time period resulted in his ideas being disregarded. Liebig, in 1872, extracted soils he obtained from the Rothamsted Experiment Station with dilute solutions of HCl, HOAc, and HNO_3 and demonstrated varying quantities of P and K were dissolved depending on the acid used. Hilgard experimented with U.S. soils in the late 1800s and promoted the use of strong HCl solutions to evaluate soil fertility status. Finally, Dyer (1894) used 1% citric acid (10 g citric acid in 1 L of water) as an extractant for P and K based on the hypothesis that it was similar in strength to root sap. He extracted soils from the Rothamsted Experiment Station and recommended phosphatic manure was needed if <100 mg kg^{-1} of phosphoric acid was extracted from the soil.

The second period described by Anderson (1960) was from 1906 to 1925. Soil chemists emphasized the relationship between the fundamental chemical composition of soils and crop production, rather than focusing on the development of rapid soil test methods. A large database of soil chemical information was compiled in many states and regions during this period, and the chemical characteristics of many soils were described. This information later became useful in improving soil test methods and interpreting soil test results. Also, during this time, Mitscherlich developed the relationship of diminishing returns between nutrient supply and plant growth, and C.G. Hopkins promoted monitoring of soil fertility status to avoid nutrient depletion.

The third period described by Anderson (1960) was from 1925 to the 1950s. Development of soil test methods were once again emphasized, stimulated in large part by increased crop yield potentials following the introduction of improved crop varieties, improved fertilizers, and increased fertilizer availability. Soil testing followed two paths during this time. Some researchers attempted to develop *universal* or multi-element extractants (i.e., solutions that extract several nutrients at once), while others attempted to adapt or develop extractants for specific nutrients under specific soil conditions. In the 1930s, Chapman and Kelley (1930) used neutral 1 M NH_4OAc as an extractant for K and other exchangeable bases. In the 1940s, Morgan (1941) developed the universal Morgan extractant using acetic acid and sodium acetate at pH 4.8, while Bray and Kurtz (1945) were developing specific tests for P in midwest soils using mixtures of NH_4F and HCl (Bray P1 and P2 methods). In the 1950s, Mehlich (1953) developed a multi-element extractant using H_2SO_4 and HCl (double acid method), while Olsen et al. (1954) developed a P extractant using pH 8.5 $NaHCO_3$. Olsen's original intent was to use the extractant for all soils (alkaline, neutral, and acid), but it gained acceptance for only alkaline soils.

The trends of Anderson's third period continued into the 1960s, 1970s, and 1980s except multi-element extractants received more attention. Multi-element extractants developed during this period included the Modified Morgan extractant (McIntosh, 1969), Soltanpour's ammonium bicarbonate-DTPA extractant (Soltanpour & Schwab, 1977), the Mississippi extractant (Lancaster, 1980), and the Mehlich 3 extractant (Mehlich, 1984). Along the way, some extractants have fallen by the wayside (e.g., carbonated water, Truog's $H_2SO_4/(NH_4)_2SO_4$ extractant, and Mehlich 2), while others have maintained their viability for >60 yr (e.g., neutral 1 M NH_4OAc). Numerous single-element and multi-element extractants resulted from research during the 1930s to 1980s, with limited agreement as to which extractants were best for specific circumstances. Less agreement was found as to which extractants could be successfully adopted as nationally standardized methods, even though the idea had been discussed by Jones (1973) and others through the years.

Ensuing from soil test development during the 1930s to 1980s are two trends that have moved to the forefront and are defining a new era in soil testing for the 1980s and 1990s. The first has been increased efforts by universities, regional soil testing groups, and the Council on Soil Testing and Plant Analysis to standardize chemical extraction techniques and adopt nationally accepted procedures. Later sections of this chapter will focus on this trend and identify selected soil test procedures for N, P, and K that are suitable for standardization and feasible for adoption nationwide. The second trend is increased investigation of soil testing approaches other than traditional chemical extraction methods. Other chapters in this monograph will discuss the trend toward alternative approaches in detail and will not be covered here. It is interesting to note, however, that soil testing is currently in a state of flux and the next 10 yr will probably see many changes in soil testing approaches at the national level. Initial changes will probably be towards standardization of extraction methods. Development of alternative approaches may occur later if advancing technology can provide significant improvements over the successes of chemical extraction techniques currently used. At the present time, no sound alternatives exist.

CHARACTERISTICS OF GOOD SOIL TEST METHODS

A prerequisite to changing an accepted soil test procedure is that the new methodology must provide improvement over the current method. A proper evaluation requires an understanding of the characteristics of a good soil test. Several authors have described good soil test methods (Bray, 1948; Viets & Lindsay, 1973; Havlin & Soltanpour, 1981). These criteria have remained essentially unchanged even though technology has advanced.

Basically, a good soil test method is correlated with crop response, reproducible, rapid, and inexpensive. Following this criteria, the *ideal* laboratory soil test would take just a few minutes to perform, the cost would be low, sample results would be the same within and across laboratories, and test results would correspond closely with crop response. Technology, of course, places limits on having an ideal test. Realistically, we must design soil tests to strike balance among the above characteristics, while striving to optimize each. In some cases, enhancing one characteristic is done at the expense of another.

A good example is the practice of scooping rather than weighing soil samples for analysis. In essence, precision is sacrificed for speed. Weighing soil samples introduces less variability from sample to sample and technician to technician. Weighing also eliminates variability associated with scoop size and shape, and simplifies standardization of procedures across laboratories. Scooping, on the other hand, is much faster. So much faster, most are willing to accept loss in precision and reproducibility for increased speed. In time, technological improvements such as automation and robotics may make weighing more feasible; but until then scooping is preferred. Another example is the choice between single-element and multi-element extractants. Multi-element extractants may not correlate with crop responses as well as single-element extractants; but because they are more cost effective, they have often been adopted. New soil test methods and approaches usually enhance some characteristics and compromise others. Their overall effect must be considered before modifying or replacing an existing method.

Soil test methods must be rapid for several reasons. Most importantly, users expect and demand fast turn-around. Sometimes this demand is due to a narrow window of opportunity for timely testing, sometimes it is due to poor planning. Regardless, laboratories will resist adopting soil tests requiring lengthy procedures.

Long turn-around times may be acceptable for environmental testing since results are often used for monitoring, making long-term management decisions, or litigation. Soil testing, however, is normally used to develop shorter term fertility programs or determining immediate fertilizer needs, provoking a need for fast turn-around. The seasonal nature of soil testing also necessitates rapid methods. Laboratories typically receive one-third to two-thirds of their annual sample volume within a 2- to 3-mo period. Soil test methods must therefore be adaptable to batch processing, procedural steps must be minimal, and analytical measurements need to be in a detection range that facilitates quick and easy determination with few interferences.

Soil tests should be inexpensive in order to attract users. Convincing farmers or fertilizer dealers, who sometimes subsidize testing services for farmers, to soil test regularly is often difficult since net returns are not always easily recognizable, especially when low value crops are grown. Additionally, when consideration is given to the time and expense involved in properly collecting samples, it is not surprising that numbers of samples received by laboratories are greatly affected by testing cost. In Oklahoma, for example, a free soil test was offered by the Oklahoma State University laboratory in the summer of 1985 between wheat harvest and fall planting of the new crop. Samples received that year increased ≈145% above the 4-yr average. Other states have reported reductions in sample numbers after increasing fees or after implementing fees for previously free testing services, further demonstrating the effect of cost on utilization of soil testing. The value of well-managed soil testing programs must be promoted more effectively to reduce large fluctuations in testing activity with small changes in price. Proper education will foster regular use of soil testing with less dependence on cost, as long as per sample costs are reasonable. Even so, cost will always be a factor. New methods or approaches to soil testing will not be accepted unless they are competitive with current approaches.

Reproducibility is another essential characteristic of good soil test methods. It pertains to the precision of a test and the degrees of variability associated with performing the test within and among laboratories. A soil test must be reproducible in order for the results to be useful and interpretable. Monitoring reproducibility provides laboratories a means of assessing quality assurance and quality control and helps establish credibility in the mind of users. Standardization of procedures is probably the best way to insure adequate reproducibility of soil test methods among laboratories, and also help counter the common complaint that soil test results vary too much from laboratory to laboratory. From a quality assurance and quality control standpoint, precision of a soil test procedure need only be slightly better than the precision of the fertilizer application method or slightly better than the variability within the soil sampling unit represented by the test. From this perspective, high levels of precision are not always needed for soil tests. Care must be taken not to overemphasize reproducibility at the expense of other criteria when assessing soil test methods.

Finally, soil test methods must be correlated with crop response to be of value. A properly correlated soil test is both accurate and interpretable. Accurate in the sense that it will measure the targeted nutrient fraction in the soil, and interpretable in the sense the results can be calibrated to determine fertilizer needs for particular soils and crops and subsequently used to predict yield response from specific fertilization rates. The ability to predict yield response allows farmers to quantify returns from fertilization and evaluate opportunity costs associated with allocating resources for other purposes. Several statistical techniques are used to estimate critical values and predict yield response (e.g., linear-plateau, quadratic, Cate-Nelson). The different techniques often result in quite different interpretations, and arguments can be made for preferring one approach over another. If the soil test method is not intrinsically related with plant response, however, no statistical technique will make the test interpretable.

FACTORS DRIVING CURRENT APPROACHES

Several factors are driving current approaches to soil testing. They can be divided into three general categories: use of new technology, attempts to increase efficiency, and efforts to improve credibility.

Technological advances have provided many opportunities to improve soil testing, and more advances are forthcoming. Development of new instrumentation, such as inductively coupled plasma spectrophotometry and ion chromatography allow simultaneous analysis of several elements, reducing analysis time and encouraging the adoption of multi-element extractants. The development of automatic samplers and refinement of flow injection analysis techniques are mechanical advances that have increased productivity of technicians and improved laboratory efficiency. Computerization and other electronic advances have made data acquisition and data processing faster and reduced errors associated with data entry. Bar-coding is used in some laboratories to speed the handling of samples and reduce errors (Allen, 1992). Future development and improvement of robotic systems for sample weighing, grinding, mixing, extracting, and diluting will improve sample processing.

The desire to improve laboratory operation efficiency is another driving force for change in soil testing. Adoption of multi-element extractants not only takes advantage of new instrumentation, but also reduces chemical costs and labor requirements. The development of new extractants, and the modification of existing ones, has often been driven by the desire to use safer, less expensive chemicals. During the development of methods, considerations are given to the effects of chemicals and materials on technician safety and health, on equipment and instrumentation life, and on what compounds will enter the laboratory waste stream. An example is the modification of the Mehlich 2 extractant to Mehlich 3 (Mehlich, 1984). Mehlich replaced NH_4Cl and HCl in the Mehlich 2 extractant with NH_4NO_3 and HNO_3 in the Mehlich 3 extractant to reduce the corrosive effect of Cl. Safety and environmental considerations are an important driving force due to increased monitoring and regulation of chemicals and waste streams, and a focus on risk assessment.

The desire to improve credibility of soil testing has stimulated efforts to develop national standardized soil testing procedures, and has stimulated the search for approaches better than chemical extraction methods. Both activities are valuable because they will bolster the credibility of soil testing as a management tool and verify the scientific basis behind it. Presently, chemical extraction techniques maintain their status as the most valid approach. Some alternative methods show promise but sufficient scientific backing has not accumulated to warrant a major change or complete switch from the chemical extraction approach. As a result, standardization of current methods will be pursued vigorously in the near term.

A first step in standardization of methods is to discourage the use of extractants that are restricted to one or very few states. A second step is to narrow the field of extractants for a particular nutrient to three or four that are broadly acceptable. Selecting multi-element extractants over single-element extractants is preferred at this stage, but not necessary. Finally, state research institutions that do not have calibration data for one of the selected extractants would be encouraged to begin work to calibrate and adopt an approved method. Areas where extractants appear to inadequately correlate with crop response should incorporate other soil characteristics into the calibration and interpretation to improve correlation to an acceptable level. In the mean time, research and development of new approaches should continue. These approaches should not be adopted, however, until a consensus is reached indicating the new approach is an improvement over the status quo.

CURRENT AND ALTERNATIVE APPROACHES TO NITROGEN TESTING

Soil N testing typically consists of one of two general approaches: (i) inorganic N tests, and (ii) N availability indexes. Inorganic N tests are the most widely used and most frequently measure residual soil NO_3–N. Availability indexes are being attempted in some areas to measure or predict N mineralization potential of a soil. Additionally, cropping and management practices are used to aid in the interpretation of test results. Recent reviews of these topics are provided by Keeney (1982), Stanford (1982), and Dahnke and Johnson (1990).

Residual inorganic N tests primarily consist of extraction and analysis of NO_3–N. In a few cases, soil testing laboratories also offer a soil NH_4–N test. Numerous studies have shown residual NO_3–N tests to be helpful in making N fertilizer recommendations, especially in relatively low rainfall regions such as the Great Plains (Dahnke & Johnson, 1990). Fertilizer requirements are credited for the amount of NO_3–N present in the soil. More recently NO_3–N soil tests have been developed in higher rainfall regions of the USA Magdoff et al. (1984) developed a soil NO_3–N test for corn (*Zea mays* L.) in Vermont. They recommended measuring soil NO_3–N in the 0- to 30-cm soil depth increment when plants were 15- to 30-cm tall to determine the need for sidedress N applications. Development of NO_3–N soil tests is expected to continue, in a large extent due to environmental concerns regarding water quality and the need to increase N fertilizer use efficiency.

In contrast to NO_3–N soil tests, minimal success has been encountered in developing N availability index tests rapid enough for routine use by soil testing laboratories. In general, chemical measures of N availability or mineralization potential take several days to weeks under controlled conditions to obtain satisfactory results. Biological tests take even longer. A third approach used in Europe

Table 12–2. Nitrate extractants currently in use by state university soil testing laboratories in the continental USA.†

Sulfate salts	Chloride salts	Other extractions
Aluminum sulfate (3)	Calcium chloride (3)	Acetic acid/sodium acetate (Morgan) (2)
Ammonium sulfate (3)	Potassium chloride (10)	Ammonium bicarbonate-DTPA (1)
Calcium sulfate (5)	Sodium chloride (1)	Calcium hydroxide (2)
Copper sulfate (1)		Calcium Oxide (2)
		Water (8)

†Number of states using the extractant is shown in parenthesis following each extractant. Seven states do not offer a NO_3 test.

is electroultrafiltration (Nemeth, 1979). Electroultrafiltration is faster than chemical and biological methods, but is relatively expensive and requires a high level of technical expertise. A more feasible approach is to estimate N availability from known or easily determined soil properties. Buchholz et al. (1981) used organic matter, cation-exchange capacity, texture, and crop type (i.e., cool vs. warm-season crop) to estimate available N through a growing season. Organic matter is most commonly used to provide a soil N availability index (Dahnke & Johnson, 1990).

Currently, 41 university soil testing laboratories offer a NO_3–N soil test either routinely or upon request. Twelve different extractants with a broad range in ionic strength are used (Table 12–2). Sulfate-salts, Cl-salts, and water are used by most of the states with KCl, water, and $CaSO_4$ ranking as the top three extractants. Calcium hydroxide, Ca oxide, and two universal extractants account for the remainder. Because soil NO_3–N is primarily in a soluble form, all the extractants remove relatively the same fraction, reducing the need to have a nationally standardized NO_3–N test. The salts in the extractant are added for various reasons, most commonly to prevent dispersion of clay and improve filtration of the extract. A high molarity salt, usually 1 to 2 M KCl, is used to remove exchangeable NH_4–N when soil NH_4–N is being measured along with NO_3–N. Extractants containing dilute salts also remove any sorbed NO_3–N which may be present in some soils containing anion exchange sites. In most instances, however, quantities of NO_3–N removed by water and salt extractants are not significantly different.

Inorganic N extractants that could most easily be adopted as standard methods include KCl, water, $CaSO_4$, and the two universal extractants, Morgan (Morgan, 1941) and Soltanpour (Soltanpour & Schwab, 1977). Only KCl and the Morgan extractant can be used for an NH_4–N test. The extractant, Soltanpour, contains NH_4, while water and saturated $CaSO_4$ are inadequate for removing exchangeable NH_4.

CURRENT AND ALTERNATIVE APPROACHES TO PHOSPHORUS TESTING

Phosphorus is present in the soil in a variety of organic and inorganic forms. This diversity has stimulated extensive research on P soil testing and has resulted in many approaches. A good review of P soil testing is provided by Fixen and Grove (1990) and Thomas and Peaslee (1973). Currently accepted approaches for

P soil testing involve conventional chemical extractions. Many alternative approaches have been and continue to be investigated.

Examples of chemical extractant alternatives include the use of anion exchange-resins, isotopic exchange, iron oxide coated strips (Menon et al., 1989), quantity intensity equilibrations (Kovar & Barber, 1988), and modeling (Fixen & Grove, 1990). These approaches are discussed in detail in another chapter. To date, they have proven too complex, expensive, or slow for implementation as routine soil test procedures. As a result, almost all soil samples are tested for P using chemical extractions.

Chemical extractions have proven successful for testing soil P. The problem is the large variety of extractants in use (Table 12–3). Many P extractants have been discarded through the years as well. When university extractants are plotted state by state on a U.S. map (Fig. 12–1), it can be seen that Bray P1 dominates the midwest, Mehlich 1 dominates the southeast, and Olsen dominates the west. Mehlich 3, Morgan, and Modified Morgan extractants are used to a lesser extent. Mehlich 3 is used in both the south central and eastern USA, while Morgan is used in New York, Massachusetts, Washington, and Idaho. Three universities

Table 12–3. Selected characteristics of P extractants currently in use by state university soil testing laboratories.

Extractant†	Composition	Suitability	Reference
Bray P1 (11)	$NH_4F + HCl$	acid soils, low to high cation-exchange capacity; neutral, noncalcareous soils, low to moderate cation-exchange capacity	Bray & Kurtz, 1945
Bray P2 (1)	$NH_4F + HCl$	acid soils, low to high cation-exchange capacity; neutral, nocalcareous soils, low to moderate cation-exchange capacity	Bray & Kurtz, 1945
Mehlich 1 (10)	$HCl + H_2SO_4$	sandy soils, acid, low cation-exchange capacity	Mehlich, 1953
Mehlich 3 (7)	$NH_4F + NH_4NO_3 +$ HOAc + HNO_3 + EDTA	wide range of soils	Mehlich, 1984
Mississippi (1)	Two step extraction: First - HCl Second - $NH_4F + AlCl_3$ + HOAc + $CH_2(COOH)_2$ + $CH_2CHOH(COOH)_2$	wide range of soils	Lancaster, 1980
Morgan (4)	HOAc + NaOAc, pH 4.8	acid soils, low to moderate cation-exchange capacity	Morgan, 1941
Modified Morgan (3)	HOAc + NH_4OAc, pH 4.8	acid soils, low to moderate cation-exchange capacity	McIntosh, 1969
Olsen (9)	$NaHCO_3$, pH 8.5	neutral and alkaline soils	Olsen et al., 1954
Soltanpour (1) (AB-DTPA)	NH_4HCO_3 + DTPA, pH 7.5	neutral and alkaline soils	Soltanpour & Schwab, 1977
Texas (1)	$NH_4OAc + HCl$ + EDTA, pH 4.2	wide range of soils	Johnson et al., 1984

†Number of states using the extractant is shown in parenthesis following each.

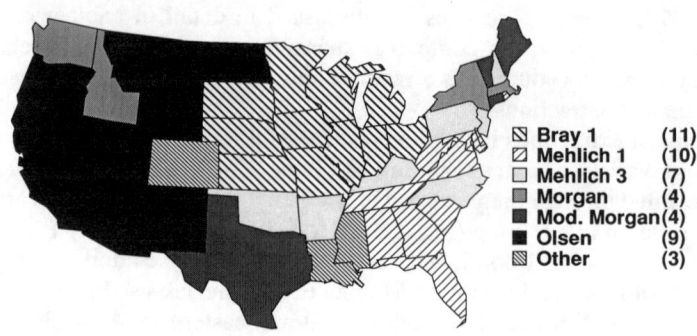

Fig. 12–1. Phosphorus extracting solutions currently used by state university soil testing laboratories in the continental USA. The number of states using each extractant is shown in parentheses in the legend.

(Maine, Vermont, and Connecticut) use the McIntosh (1969) version of the Modified Morgan, while Texas has its own version of the Modified Morgan (Table 12–3). The three remaining universities have their own unique extractants (i.e., Louisiana uses Bray P2, Mississippi uses Lancaster, and Colorado uses Soltanpour).

The Mehlich 3 extractant is the newest (Mehlich, 1984), and is increasing in popularity with five more states planning to adopt Mehlich 3 by 1996 (Delaware, Louisiana, Maryland, Virginia, and West Virginia) (Fig. 12–2). In addition, Connecticut, Florida, and Texas are considering adoption of Mehlich 3, and the North Central Region NCR-13 Soil and Plant Analysis Committee has approved it as an alternative extractant to Bray P1 (Munter & Schulte, 1993). If the above potential changes are implemented, the distribution of P extractants across the USA will look as shown in Fig. 12–3.

If Mississippi and the remaining Mehlich 1 users converted to Mehlich 3, Colorado switched to Olsen, and the six remaining users of Morgan or Modified Morgan came to a consensus on a New Morgan extractant, the entire continental USA could be represented by three P extractants that could be adopted as nationally approved standard methods (Fig. 12–4). Two of the three extractants in this scenario are multi-element extractants and extract both P and K. If three multi-element extractants were desired, an alternative scenario is to replace the Olsen

SOIL TESTING METHODS

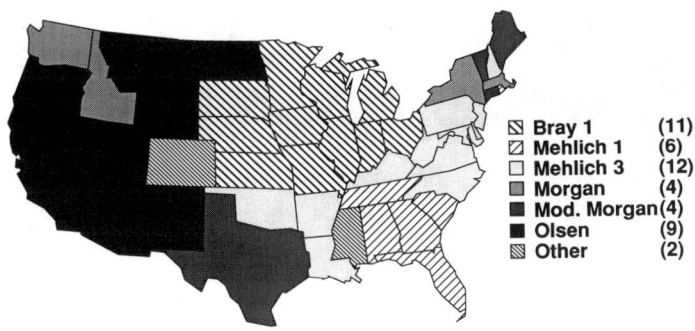

Fig. 12–2. Phosphorus extracting solutions planned to be in use by state university soil testing laboratories in 1996 in the continental USA. The number of states for each extractant is shown in parentheses in the legend.

method with the Soltanpour method assuming the western states could agree to the change. Realistically this approach would be more difficult and not necessarily an improvement since many of the western states rarely use K tests anyway.

The above discussion is a realistic and reasonable approach for identifying standard procedures for P. The Olsen method is a good choice based on its current widespread use in the west. Mehlich 3 is attractive based on its current popularity and the trend toward its adoption in the central, southern, and eastern areas of the USA. The Morgan method is justified for two reasons. First, it has been shown in the northeast and northwest to perform better than Mehlich and Bray methods, especially on medium testing soils. Convincing users of Morgan extractants to switch to Mehlich 3 or Olsen extractants may be undesirable and difficult. Secondly, the Morgan extractant is similar to the extractant defined by the Environmental Protection Agency for their Toxicity Characteristic Leaching Procedure (TCLP). The TCLP was designed to determine the mobility of inorganic and organic contaminants in liquid, solid, and multiphasic wastes (U.S. Environmental Protection Agency, 1990). Standardizing an extractant similar to one already regulated may prove beneficial as soil scientists increase their role in guiding U.S. Environmental Protection Agency and other governmental agencies into making science-based regulatory decisions. Overall, narrowing the number of P extractants in the USA from 10 to 3 represents a substantial

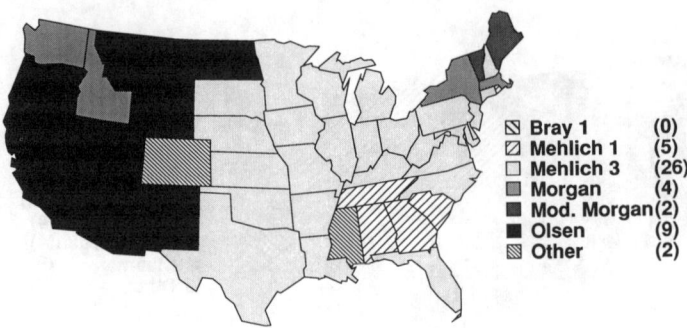

Fig. 12–3. Phosphorus extracting solutions expected to be used by state university soil testing laboratories in the continental USA after 1996 if changes under consideration are implemented. The number of states for each extractant is shown in parentheses in the legend.

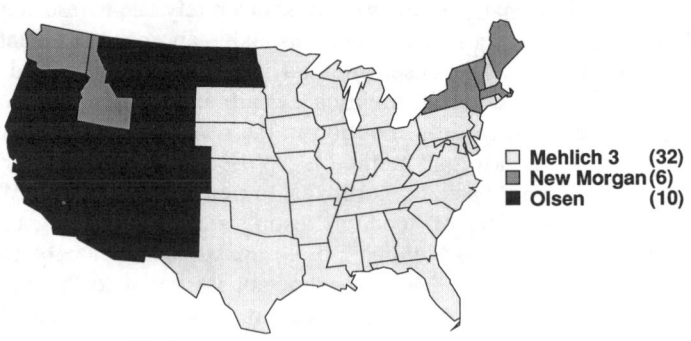

Fig. 12–4. Phosphorus extraction procedures recommended for adoption as standard methods in the continental USA. The New Morgan extractant would be determined by states currently using the Morgan and Modified Morgan extractants.

improvement over the status quo and warrants strong consideration. Such a move would certainly go far towards improving public perception of soil testing reliability.

CURRENT AND ALTERNATIVE APPROACHES TO POTASSIUM TESTING

Exchangeable K has been found to be correlated with plant uptake and crop response under numerous conditions (Doll & Lucas, 1973; Haby et al., 1990). As a result, extraction of exchangeable K, which includes water soluble K, has been the dominate approach to K soil testing for many years. Even so, alternative approaches exist and continue to be evaluated. In this section, traditional and alternative approaches to K soil testing will be briefly discussed, followed by the identification and discussion of K soil test extractants suitable for adoption as national standards.

The current approach to soil K testing uses many of the same multi-element and single element extractants used to extract P (Table 12–4). Of the 11 extractants currently used by state university soil testing laboratories, nine are also used to extract P. Two of the nine, Bray P1 and Olsen, are usually single-element extractants, but have been adapted for K by two states (Wisconsin and Utah, respectively). Neutral NH_4OAc and water are the only K extractants that also are not used for P extraction. Neutral NH_4OAc extracts exchangeable K and is used by 17 states. Water extracts soluble K and is only used by New Mexico.

Alternative approaches for K include the use of cation exchange resins (Skogley & Georgitis, 1988), quantity-intensity equilibrium calculations, Baker's quantity-intensity extraction-incubation procedure (Baker & Amacher, 1981), and electroultrafiltration (Nemeth, 1979). These approaches are discussed in more detail in another chapter. Currently, they are too complex, expensive, or time consuming to be used routinely.

Potassium extractants in current use by state university soil testing laboratories in the continental USA are illustrated in Fig. 12–5. Seventeen states use neutral $1 \, M \, NH_4OAc$, primarily in the north central and western USA. The distribution of Lancaster, Mehlich 1, Mehlich 3, Morgan, and Modified Morgan extractants are similar to P since the states involved use these extractants for both P and K. The Soltanpour extractant is used by two states, Colorado and Wyoming. Utah uses a modified Olsen, New Mexico uses water, and Wisconsin uses Bray P1. Mehlich 3 is gaining in popularity, and by 1996 will have been adopted by five more states (Delaware, Louisiana, Maryland, Virginia, and West Virginia), changing the distribution to that shown in Fig. 12–6. In addition, Connecticut, Florida,

Table 12–4. Potassium extractants currently in use by state university soil testing laboratories.†

Bray P1 (1)	Morgan (4)	Soltanpour (AB-DTPA) (2)
Mehlich 1 (10)	Modified Morgan (3)	Texas (1)
Mehlich 3 (7)	Neutral ammonium acetate (17)	Water (1)
Mississippi (1)	Olsen (1)	

†Number of states using the extractant is shown in parenthesis following each.

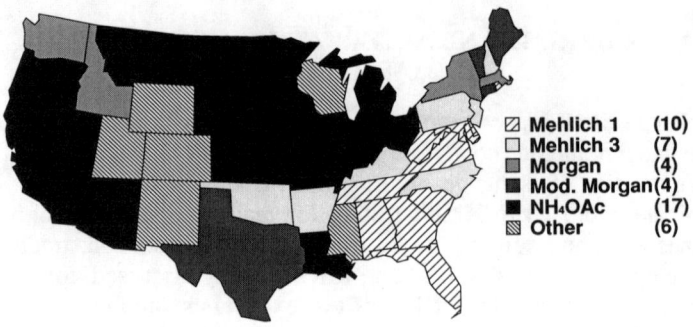

Fig. 12–5. Potassium extracting solutions currently used by state university soil testing laboratories in the continental USA. The number of states using each extractant is shown in parentheses in the legend.

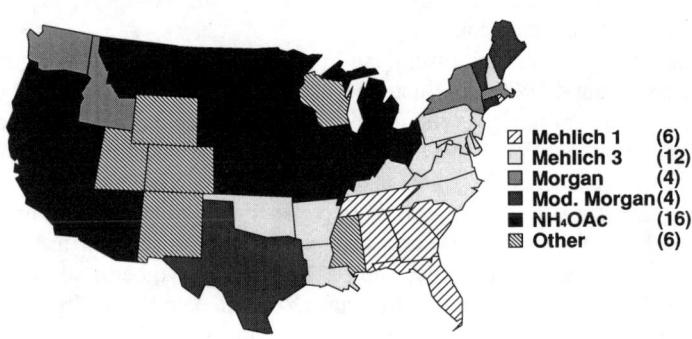

Fig. 12–6. Potassium extracting solutions planned to be in use by state university soil testing laboratories in 1996 in the continental USA. The number of states for each extractant is shown in parentheses in the legend.

and Texas are considering adopting Mehlich 3, and the North Central Region NCR-13 Soil and Plant Analysis Committee has approved it as an alternative extractant to NH_4OAc (Munter & Schulte, 1993). New Mexico offers neutral NH_4OAc upon request.

Implementation of these potential changes leaves Mehlich 3 as the dominant extract with others restricted to selected smaller regions of the USA (Fig. 12–7). Following the same argument as for P, three standard extractants (Mehlich 3, New Morgan, and NH_4OAc) covering the entire USA could be established by making three broad changes: Mississippi and the Mehlich 1 states converting to Mehlich 3, Colorado, Utah, and Wyoming switching to NH_4OAc, and states using Morgan and Modified Morgan solutions uniting behind a uniform Morgan procedure (Fig. 12–8). At least some of these changes are forthcoming in the near future.

The ideas suggested here, for both P and K are reasonable, but require the cooperation, time, and expense that comes with change. In the end, standardization of procedures will be worth the effort due to increased credibility and more extensive use of soil testing.

SUMMARY

Current technology dictates that rapid chemical extraction is still the most feasible approach to soil testing. A reduction in the diversity of extractants and

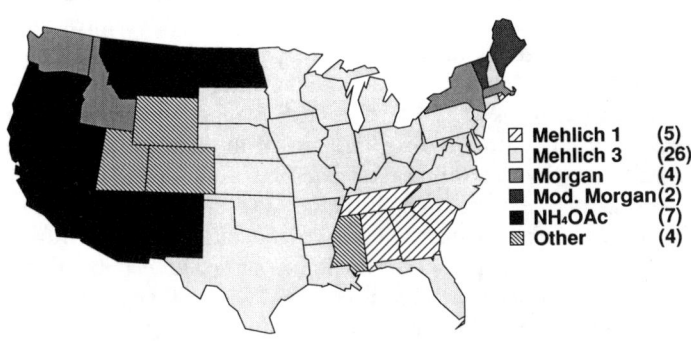

Fig. 12–7. Potassium extracting solutions expected to be used by state university soil testing laboratories in the continental USA after 1996 if changes under consideration are implemented. The number of states for each extractant is shown in parentheses in the legend.

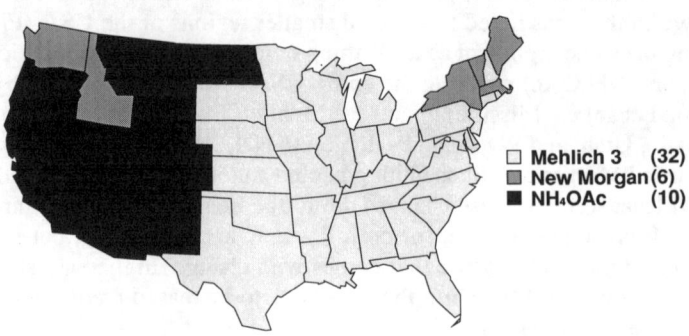

Fig. 12–8. Potassium extraction procedures recommended for adoption as standard methods in the continental USA. The New Morgan extractant would be determined by states currently using the Morgan and Modified Morgan extractants.

adoption of standard methods is desirable. A stepwise and realistic approach to national standardization of soil test methods for N, P, and K was presented, resulting in three standard methods for P (Mehlich 3, New Morgan, and Olsen), three standard methods for K (Mehlich 3, New Morgan, and NH_4OAc), and five standard methods for inorganic N ($CaSO_4$, KCl, Morgan, Soltanpour, and water). States having difficulty correlating crop response with one of the standard extractants would be encouraged to modify and refine their interpretations taking into account soil properties such as pH, texture, organic matter, and cation-exchange capacity to improve the correlation rather than rejecting a standard procedure. Research and development of alternative approaches should be continued, but adoption of a new approach should occur only if a consensus agrees the new approach represents a substantial improvement based on the criteria outlined for a good soil test.

REFERENCES

Allen, E.R., and T. Ong. 1992. Computer bar-coding system for data acquisition in a soil and plant analytical laboratory. p. 25. *In* Agronomy abstracts. ASA, Madison, WI.

Anderson, M.S. 1960. History and development of soil testing. J. Agric. Food Chem. 8:84–87.

Baker, D.E., and M.C. Amacher. 1981. The development and interpretation of a diagnostic soil testing program. Pennsylvania State Univ. Agric. Exp. Stn. Bull. 826. University Park.

Bray, R.H. 1948. Requirements for successful soil tests. Soil Sci. 66:83–89.

Bray, R.H., and L.T. Kurtz. 1945. Determination of total, organic, and available forms of phosphorus in soils. Soil Sci. 59:39–45.

Buchholz, D.D, J.R. Brown, R.G. Hanson, H.N. Wheaton, and J.D. Garrett. 1981. Soil test interpretations and recommendations handbook. Univ. of Missouri, Columbia.

Cate, R.B., Jr., and L.A. Nelson. 1965. A rapid method for correlation of soil test analyses with plant response data. North Carolina State Univ. Int. Soil Testing Ser. Tech. Bull. 1. Raleigh.

Chapman, H.D., and W.P. Kelley. 1930. The determination of the replaceable bases and base exchange capacity of soils. Soil Sci. 30:391–406.

Dahnke, W.C., and G.V. Johnson. 1990. Testing soils for available nitrogen. p. 127–139. *In* R.L. Westerman (ed.) Soil testing and plant analysis. 3rd ed. SSSA Book Ser. 3. SSSA, Madison, WI.

Daubeny, C. 1845. VII Memoir on the rotation of crops and on the quantity of inorganic matter abstracted from the soil by various plants under different circumstances. R. Soc. London Philos. Trans. 135:179–253.

Doll, E.C., and R.E. Lucas. 1973. Testing soils for potassium, calcium, and magnesium. p. 133–151. *In* L.M. Walsh and J.D. Beaton (ed.) Soil testing and plant analysis. SSSA, Madison, WI.

Dyer, B.J. 1894. On the analytical determination of probably available mineral plant food in soils. J. Chem. Soc. (London). 65:115–167.

Fixen, P.E., and J.H. Grove. 1990. Testing soils for phosphorus. p. 141–180. *In* R.L. Westerman (ed.) Soil testing and plant analysis. 3rd ed. SSSA Book Ser. 3. SSSA, Madison, WI.

Haby, V.A., M.P. Russelle, and E.O. Skogley. 1990. Testing soils for potassium, calcium, and magnesium. p. 181–227. *In* R.L. Westerman (ed.) Soil testing and plant analysis. 3rd ed. SSSA Book Ser. 3. SSSA, Madison, WI.

Havlin, J.L., and P.N. Soltanpour. 1981. Evaluation of the NH_4HCO_3-DTPA soil test for iron and zinc. Soil Sci. Soc. Am. J. 45:70–75.

Johnson, G.V., R.A. Isaac, S.J. Donohue, M.R. Tucker, and J. Woodruff. 1984. Procedures used by state soil testing laboratories in the southern region of the United States. Agric. Exp. Stn. Southern Coop. Ser. Bull. 190. Oklahoma State Univ., Stillwater.

Jones, J.B. 1973. Soil testing in the United States. Commun. Soil Sci. Plant Anal. 4:307–322.

Keeney, D.R. 1982. Nitrogen-availability indices. p. 711–733. *In* A.L. Page et al. (ed.) Methods of soil analysis. Part 2. 2nd ed. Agron. Monogr. 9. ASA and SSSA, Madison, WI.

Kovar, J.L., and S.A. Barber. 1988. Phosphorus supply characteristics of 33 soils as influenced by seven rates of phosphorus addition. Soil Sci. Soc. Am. J. 52:160–165.

Lancaster, J.D. 1980. Mississippi soil test methods and interpretation. Mississippi Agric. Exp. Stn. Mimeo. Mississippi State Univ., Mississippi State.

Magdoff, F.R., D. Ross, and J. Amadon. 1984. A soil test for nitrogen availability in corn. Soil Sci. Soc. Am. J. 48:1301–1304.

McIntosh, J.L. 1969. Bray and Morgan soil extractants modified for testing acid soils from different parent materials. Agron. J. 61:259–265.

Mehlich, A. 1953. Determination of P, K, Na, Ca, Mg, and NH_4. Soil Test Div. Mimeo. North Carolina Dep. Agric., Raleigh.

Mehlich, A. 1984. Mehlich 3 soil test extractant: A modification of the Mehlich 2 extractant. Commun. Soil Sci. Plant Anal. 15:1409–1416.

Menon, R.G., L.L. Hammond, and H.A. Sissingh. 1989. Determination of plant-available phosphorus by the P_i soil test. Soil Sci. Soc. Am. J. 53:110–115.

Morgan, M.F. 1941. Chemical soil diagnosis by the universal soil testing system. Connecticut Agric. Exp. Stn. Bull. 450. Univ. of Connecticut, Storrs.

Munter, R.C., and E.E. Schulte. 1993. An NCR-13 summary of Mehlich 3 research. R.C. Munter, Dep. of Soil Sci, Univ. of Minnesota, Minneapolis.

Nemeth, K. 1979. The availability of nutrients in the soil as determined by electro-ultrafiltration (EUF). Adv. Agron. 31:155–188.

Olsen, S.R., C.V. Cole, F.S. Watanabe, and L.A. Dean. 1954. Estimation of available phosphorus in soils by extraction with sodium bicarbonate. USDA Circ. 939. U.S. Gov. Print. Office. Washington, DC.

Skogley, E.O., and S.J. Georgitis. 1988. Phytoavailability soil test. p. 164–167. *In* J.L. Havlin (ed.) Proc. Great Plains Soil Fert. Workshop, Denver, CO. 8–9 Mar. Kansas State Univ., Manhattan, KS.

Soltanpour, P.N., and A.P. Schwab. 1977. A new soil test for simultaneous extraction of macro- and micro-nutrients in alkaline soils. Commun. Soil Sci. Plant Anal. 8:195–207.

Stanford, G. 1982. Assessment of soil nitrogen availability. p. 651–688. *In* F.J. Stevenson (ed.) Nitrogen in agricultural soils. Agron. Monogr. 22. ASA, CSSA, and SSSA, Madison, WI.

Thomas, G.W., and D.E. Peaslee. 1973. Testing soils for phosphorus. p. 115–132. *In* L.M. Walsh and J.D. Beaton (ed.) Soil testing and plant analysis. SSSA, Madison, WI.

Tisdale, S.L., W.L. Nelson, and J.D. Beaton. 1985. Soil fertility and fertilizers. Macmillan. New York.

U.S. Environmental Protection Agency. 1990. Hazardous waste management system; identification and listing of hazardous waste; toxicity characteristics revisions. p. 11798-11877. *In* Federal Register. U.S. Environmental Protection Agency, Washington, DC.

Viets, F.G. 1980. Present status of soil and plant analysis for fertilizer recommendations and improvement of soil fertility p. 9–20. *In* Soil and plant testing and analysis. FAO Soils Bull. 38/1. Food and Agric. Organ. of the United Nations, Rome, Italy.

Viets, F.G., and W.L. Lindsay. 1973. Testing soils for zinc, copper, manganese, and iron. p. 153–172. *In* L.M. Walsh and J.D. Beaton (ed.) Soil testing and plant analysis. SSSA, Madison, WI.